Supplements to the 2nd Edition of

RODD'S CHEMISTRY OF CARBON COMPOUNDS

ELSEVIER SCIENCE PUBLISHERS B.V.
Sara Burgerhartstraat 25
P.O. Box 211, 1000 AE Amsterdam, The Netherlands

Distributors for the United States and Canada:

ELSEVIER SCIENCE PUBLISHING COMPANY INC.
52, Vanderbilt Avenue
New York, NY 10017, U.S.A.

Library of Congress Card Number: 64-4605

ISBN 0-444-42989-1

Printed in The Netherlands

Supplements to the 2nd Edition of

RODD'S CHEMISTRY OF CARBON COMPOUNDS

VOLUME I

ALIPHATIC COMPOUNDS

★

VOLUME II

ALICYCLIC COMPOUNDS

★

VOLUME III

AROMATIC COMPOUNDS

★

VOLUME IV

HETEROCYCLIC COMPOUNDS

★

VOLUME V

MISCELLANEOUS
GENERAL INDEX

★

Supplements to the 2nd Edition (Editor S. Coffey) of

RODD'S CHEMISTRY OF CARBON COMPOUNDS

A modern comprehensive treatise

Edited by

MARTIN F. ANSELL
Ph.D., D.Sc. (London) F.R.S.C. C. Chem.
Reader Emeritus, Department of Chemistry,
Queen Mary College, University of London, Great Britain

Supplement to

VOLUME III AROMATIC COMPOUNDS

Part H:

Polycarbocyclic Compounds with more than Thirteen Atoms in the Fused-ring System

ELSEVIER
Amsterdam — Oxford — New York — Tokyo 1988

CONTRIBUTORS TO THIS VOLUME

HOWARD F. ANDREWS, B.Sc., Ph.D., C.Chem., F.R.S.C.
Department of Chemistry, Glasgow College of Technology,
Glasgow G4 0BA

ROGER BOLTON, B.Sc., Ph.D., C.Chem., F.R.S.C.
Department of Chemistry, Royal Holloway
and Bedford New College, University of London,
Surrey TW20 0EY

MALCOLM SAINSBURY, D.Sc., Ph.D., C.Chem., F.R.S.C.
Department of Chemistry, The University,
Bath BA2 7AY
(INDEX)

PREFACE TO SUPPLEMENT IIIH

This volume continues the supplementation of the second edition of Rodd's Chemistry of Carbon Compounds thus keeping this major work of reference up to date. The contents of this volume supplement chapters 28, 29 and 30 of Volume IIIH of the second edition, and thus completes the supplementation of Volume III covering Aromatic Chemistry. Roger Bolton has written the chapter (28) covering anthracene, phenanthrene and related compounds and Howard Andrews has supplemented those topics on which he wrote in the second edition namely polycyclic aromatic compounds containing one or more five membered rings (chapter 29) and polycyclic compounds with four or more (mainly 6-membered) fused carbocyclic rings (chapter 30). The index has been compiled by Malcolm Sainsbury.

At a time when there are many specialist reviews monographs and reports available, there is still in my view an important place for a book such as "Rodd" which gives a broader coverage of organic chemistry. One aspect of the value of this work is that it allows an expert in one field to quickly find out what is happening in other fields of chemistry. On the other hand a chemist looking for the way into a field of study will find within Rodd an outline of the important aspects of that area of chemistry together with leading references to other works to provide more detailed information.

This volume has been produced by direct reproduction of the authors' manuscripts. I am most grateful to both contributors for the care and effort they have put into the preparation of their manuscripts both from the point of view of their scientific content and their visual appearance. I also acknowledge the help given to me by the staff at Elsevier and thank them for guiding the transformation of the authors' manuscripts to a published work.

March 1988 Martin Ansell

CONTENTS

VOLUME III H

Aromatic Compounds: Polycarbocyclic Compounds with more than Thirteen Atoms in the Fused-ring- System

OFFICIAL PUBLICATIONS

B.P.	British (United Kingdom) Patent
F.P.	French Patent
G.P.	German Patent
Sw.P.	Swiss Patent
U.S.P.	United States Patent
U.S.S.R.P.	Russian Patent
B.I.O.S.	British Intelligence Objectives Sub-Committee Reports
F.I.A.T.	Field Information Agency, Technical Reports of U.S. Group Control Council for Germany
B.S.	British Standards Specification
A.S.T.M.	American Society for Testing and Materials
A.P.I.	American Petroleum Institute Projects
C.I.	Colour Index Number of Dyestuffs and Pigments

SCIENTIFIC JOURNALS AND PERIODICALS

With few obvious and self-explanatory modifications the abbreviations used in references to journals and periodicals comprising the extensive literature on organic chemistry, are those used in the World List of Scientific Periodicals.

LIST OF COMMON ABBREVIATIONS AND SYMBOLS USED

A	acid
Å	Ångström units
Ac	acetyl
a	axial; antarafacial
as, *asymm.*	asymmetrical
at	atmosphere
B	base
Bu	butyl
b.p.	boiling point
C, mC and μC	curie, millicurie and microcurie
c, *C*	concentration
C.D.	circular dichroism
conc.	concentrated
crit.	critical
D	Debye unit, 1×10^{-18} e.s.u.
D	dissociation energy
D	dextro-rotatory; dextro configuration
DL	optically inactive (externally compensated)
d	density
dec. or decomp.	with decomposition
deriv.	derivative
E	energy; extinction; electromeric effect; Entgegen (opposite) configuration
E1, E2	uni- and bi-molecular elimination mechanisms
E1cB	unimolecular elimination in conjugate base
e.s.r.	electron spin resonance
Et	ethyl
e	nuclear charge; equatorial
f	oscillator strength
f.p.	freezing point
G	free energy
g.l.c.	gas liquid chromatography
g	spectroscopic splitting factor, 2.0023
H	applied magnetic field; heat content
h	Planck's constant
Hz	hertz
I	spin quantum number; intensity; inductive effect
i.r.	infrared
J	coupling constant in n.m.r. spectra; joule
K	dissociation constant
kJ	kilojoule

LIST OF COMMON ABBREVIATIONS

k	Boltzmann constant; velocity constant
kcal	kilocalories
L	laevorotatory; laevo configuration
M	molecular weight; molar; mesomeric effect
Me	methyl
m	mass; mole; molecule; *meta-*
ml	millilitre
m.p.	melting point
Ms	mesyl (methanesulphonyl)
[M]	molecular rotation
N	Avogadro number; normal
nm	nanometre (10^{-9} metre)
n.m.r.	nuclear magnetic resonance
n	normal; refractive index; principal quantum number
o	*ortho-*
o.r.d.	optical rotatory dispersion
P	polarisation, probability; orbital state
Pr	propyl
Ph	phenyl
p	*para-*; orbital
p.m.r.	proton magnetic resonance
R	clockwise configuration
S	counterclockwise config.; entropy; net spin of incompleted electronic shells; orbital state
S_N1, S_N2	uni- and bi-molecular nucleophilic substitution mechanisms
S_Ni	internal nucleophilic substitution mechanisms
s	symmetrical; orbital; suprafacial
sec	secondary
soln.	solution
symm.	symmetrical
T	absolute temperature
Tosyl	*p*-toluenesulphonyl
Trityl	triphenylmethyl
t	time
temp.	temperature (in degrees centigrade)
tert.	tertiary
U	potential energy
u.v.	ultraviolet
v	velocity
Z	zusammen (together) configuration

LIST OF COMMON ABBREVIATIONS

α	optical rotation (in water unless otherwise stated)
$[\alpha]$	specific optical rotation
αA	atomic susceptibility
αE	electronic susceptibility
ε	dielectric constant; extinction coefficient
μ	microns (10^{-4} cm); dipole moment; magnetic moment
μB	Bohr magneton
μg	microgram (10^{-6}g)
λ	wavelength
ν	frequency; wave number
$\chi, \chi d, \chi\mu$	magnetic, diamagnetic and paramagnetic susceptibilities
~	about
(+)	dextrorotatory
(-)	laevorotatory
(±)	racemic
$\ominus$	negative charge
$\oplus$	positive charge

Chapter 28

ANTHRACENE, PHENANTHRENE AND RELATED COMPOUNDS

R. BOLTON

1. *Anthracene and derivatives*

Anthracene is a thermodynamically stable compound, so that it becomes a major product in the pyrolysis of a number of aromatic compounds such as xylenols (V.V. Platonov *et. al.*, Chem. Abs., 1983, 98, 71573y ; M. Janik, *ibid.*, 1983, 98, 10993a) and benzyl chlorides (L.A. Avaca, E.R. Gonzalez and E.A. Ticianelli, Electrochim. Acta, 1983, 28, 1473). Pyrolysis of phthalic esters gives anthracene (J.L. Bove and P. Dalven, Chem. Abs., 1984, 101, 42802x), whereas phthalic acids and anhydrides give biphenylenes, through benzynes, on pyrolysis (R.F.C. Brown *et. al.*, Austral. J. Chem., 1967, 20, 139). The flash vacuum pyrolysis of 2-methylbenzophenone (800-900°) gives both anthracene and fluorene; the latter prevails. However pyrolysis of the derived ketimine gives a preponderance of anthracene (T.Y. Gu and W.P. Weber, J. org. Chem., 1980, 45, 2541). The pyrolysis of 7-(dimethylsilano)-norbornadiene gives anthracene in good yield (B. Mayer and W.P. Neumann, Tetrahedron Letters, 1980, 21, 4887). Anthracene itself arises from the oxidation (30% H_2O_2) of 9,10-dihydroanthacen-9-imines (G.W. Gribble *et. al.*, *ibid.*, 1976, 3673).

Interest in the carcinogenic products of the combustion of a number of substances has led to the discovery of anthracene derivatives in the urine of human cigarette smokers (2-anthrylamine; T.H. Connor *et. al.*, Chem. Abs., 1983, 98, 138834e), among the components of sediments in Tokyo Bay (H. Takada, R. Ishiwatari and S.J. Yun, Chem. Abs., 1985, 102, 31690d), and at significant levels in rooms where tobacco has been smoked (C.K. Huynh *et. al.*, Chem. Abs., 1985, 101, 15662g) and in Finnish saunas

(E. Hasanen *et. al.*, Chem. Abs., 1985, 101, 13614h). Interest in mechanisms of carcinogensis has promoted research in this section of organic chemistry; so has the discovery of liquid crystals and their use in display devices, and the conjoint use of a range of anthracene-based dyes. In many cases, though the chemistry behind the newly synthesised compound is well-known. This Supplement will not list these products of classical reactions but will only draw attention to the application of new synthetic methods.

(a) Applications

Anthracene derivatives have been especially studied because of their fluorescent properties; compounds such as 9,10-diphenylanthracene and 9,10-di(phenylethynyl)-anthracene are used in various formulations of "lightsticks" in which an oxidation, often that of an oxalate ester induces light emission from such derivatives (D. Potrawa and A. Schlep, Chem. Abs., 1983, 99, 193921b). More imaginative uses of these properties come in the use of anthracenyl crown ethers and cryptands as fluorescent probes in biochemical systems (U. Herrmann *et. al.*, Biochemistry, 1984, 23, 4059) and in the use of 1-anthracenoyl nitrile to form fluorescent derivatives of ethanol analogues, so that they may be detected at much lower limits (0.05 - 0.2 ppm) in HPLC (M. Kudoh *et. al.*, J. Chromatogr. 1984, 287, 337). Membrane fluorescent probes have been achieved through the addition of anthracene ($AlCl_3$, CH_2Cl_2, 1 h., 20°) across olefinic bonds in phospholipids, especially those with unsaturated carbon chains attached to C-2 of the glycerol moiety (O.V. Bogomolov, A.P. Kaplan and V.I. Shvets, Bioorg. Khim., 1984, 10, 1560).

The use of optically active 2,2,2-trifluoro-1-(9-anthracenyl)ethanol has been proposed to study enantiometrically enriched alkyl substituted γ-lactones by nmr methods (W.H. Porkle, D.L. Sikkenga and M.S. Pavlin, J. org. Chem., 1977, 42, 384).

(b) Substitution

Acylation - The di-acyl derivatives of anthracene have become important both in the synthesis of materials which undergo oxidation with chemiluminesence (*e.g.* T. Hiramatsu,

T. Harada and T. Yamaji, Bull. chem. Soc. Japan, 1981, 94, 985). Aceanthrylene (1) may be obtained in a five-step process (27% over all) from anthracene, the first step being a "cycloacylation" in which oxalyl chloride adds simultaneously to C-1 and C-9 in anthracene to give the quinone (B.F. Plummer and Z. Y. Al-Saigh, J. org. Chem., 1984, 49, 2069); analogously, chloroacetyl chloride and anthracene give 1-aceanthrenone (2) from which (1) may then be obtained (H.D. Becker and L. Hansen, J. org. Chem., 1985, 50, 277).

(2) (1)

Scheme 1

The kinetics of acetylation of anthracene ($AlCl_3$, $(CH_2Cl)_2$ are first-order in both anthracene and the acylating agent (M.A. Ahmed, M.M. Elsemongy and M.F. Armira, Indian J. Chem., 1980, 19B, 414). In the presence of catalytic amounts of aluminium chloride or iron(III) chloride, p-toluoyl chloride attacks anthracene to give far more of the 9-isomer than of the 1- or of the 2-isomer, an observation in total agreement with previous experiments. On further heating in the presence of larger amounts (1.5 mole) of aluminium chloride, the relative yield of 9-isomer falls (I.K. Buchina, A.I. Bokova, and N.G. Sidorova, Chem. Abs., 1978, 88, 62202t). The 1-, 2-, and 9-acylanthracenes interconvert when heated with iron(III) chloride in phenetole (I.K. Buchina *et. al.*, Chem. Abs., 1979, 90, 168350w) but 9-acylanthracenes also give acylphenetoles and anthracenes. Among a range of 9-aroylanthracenes (Ar=Ph or $X.C_6H_4$) the total electron density of the ketone appears to determine whether acylation of the ether solvent is observed. When X = m-Cl, p-Cl, or m-MeO, solvent attack only occurs if a methyl group is present at C-10 ($FeCl_3$, 175°, PhOEt):

(A.I. Bokova and I.K. Buchina, Zh. org. Khim., 1984, 20, 1318).

Benzoylimidazole in trifluoroacetic acid also acylates anthracene at C-9 (T. Keumi, H. Saga, and H. Kitajima, Bull. chem. Soc. Japan, 1980, 53, 1638).

Photo-induced acylation of anthracene gives 2- (20-30%) and 9- (30-60%) acyl derivatives (T. Tamaki and K. Shiroki, Chem. Abs., 1979, 90, 21927f ; T. Tamaki and K. Hokoku, *ibid.*, 1984, 101, 170372u) together with small amounts of photo-dimers (T. Tamaki, Bull. chem. Soc. Japan, 1978, 51, 1145). Acylation of 9-substituted anthracenes is extensively used to produce precursors of industrially important materials, and expectedly the 10-acyl derivative predominates. The acetylation (CH_3COCl, $AlCl_3$ $(CH_2Cl)_2$) of 2-methoxyanthracene is reported to give the 1-isomer but the evidence for this conclusion is arguable. Aromatic protons, but not their relative orientations, are discernible in the ^{1}H-nmr spectrum, and the anticipated cyclodehydration with aluminium chloride was not seen (R.C. Hadden, Austral. J. Chem., 1982, 35, 1733). The expected site of substitution is C-9, and the product may in fact be the 9-acetyl-2-methoxyanthracene.

Halogenation

Fluorination - The addition of fluorine to give perfluoro-anthracene (D. Harrison, R. Stephens and J.C. Tatlow, Tetrahedron, 1963, 19, 1893) affords a route to decafluoroanthracene through defluorination by passage over hot iron(II) oxide but better yields are reported (J. Burdon *et. al.*, Chem. Comm., 1982, 534) in the sequence shown in Scheme 2. This procedure involves the synthesis of octafluoroanthraquinone by heating tetrafluorophthalic anhydride with potassium fluoride (G.G. Yakobson *et. al.*, Tetrahedron Letters, 1965, 4473; Zh. org. Khim., 1971, 7, 745). Sulphur tetrafluoride converts octafluoro-anthraquinone into 1,2,3,4,5,6,7,8,9,9,10,10-dodecafluoro-9,10-dihydroanthracene, and the removal of two atoms of fluorine gives the fully aromatic structure. Decafluoroanthracene undergoes predominant or exclusive nucleophilic attack at C-2, in contrast to the expectations of recent theoretical proposals (I.W. Parsons and J. Burdon, J. Amer. chem. Soc., 1977, 99 , 7445) although consistent with earlier explanations of the orientation of nucleophilic attack of perfluoroarenes (J. Burdon,

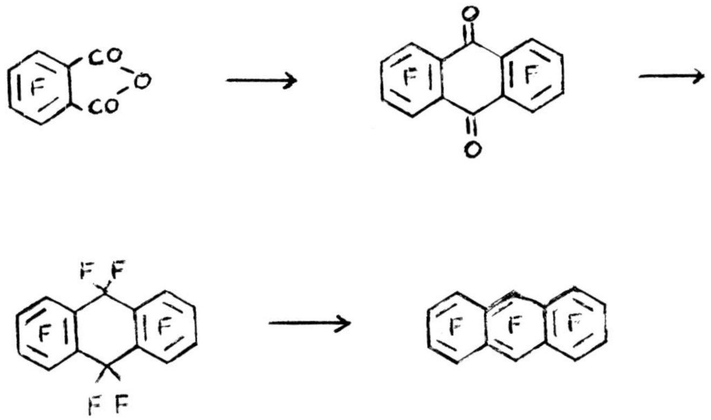

Scheme 2

Tetrahedron, 1965, 21, 3373). Mono-fluorinated anthracenes have been obtained by the conventional Balz-Schiemann process, usually carried out using the readily available aminoanthraquinones; more recently, direct methods of fluorination have been introduced. The intercalation products of xenon fluorides with graphite (*e.g.* $C_{19}XeF_6$, $C_{8.7}XeOF_4$) serve as somewhat attenuated fluorinating agents compared to their uncomplexed analogues; thus, xenon difluoride (with or without hydrogen fluoride) affords 1- (30%), 2- (4%), and 9- (36%) fluoroanthracene with anthracene, whereas $C_{19}XeF_6$ gives only 6%, 1% and 10% respectively of these three isomers. (M. Rabinovitz *et. al.*, J. chem. Res.(S), 1977, 2126. Electrochemical methods of fluorination are held (P.F. King and R.F. O'Malley, J. org. Chem., 1984, 49, 2803) to involve radical cationic intermediates. Anthracene, electrolysed with iodine and silver(I) fluoride in acetonitrile, provides 9-fluoroanthracene, but 9-nitroanthracene and phenanthrene are inert under these conditions (King and O'Malley, *loc. cit.*). Controlled potential electrolysis of anthracene, its 9-methyl or 9-phenyl derivative in solutions of $Me_4NF.2HF$ in acetonitrile gives the monofluoro (9- or 10-) product. Dimerisation limits the preparative use of this process

(J.F. Carpenter *et. al.*, J. electrochem. Soc., 1983, 130, 2170).

Chlorination - The reaction of anthracene with chlorine has not yet received the detailed qualitative study which phenanthrene has received.

The synthesis of dodecachloro-9,10-dihydroanthracene (M. Ballester *et. al.*, An. Quim., 1980, 76C, 157) by treating anthracene with Silberrad's reagent ($AlCl_3/S_2Cl_2/SO_2Cl_2$: O. Silberrad, J. chem. Soc., 1921, 119, 2029; O. Silberrad, C.A. Silberrad and B. Parke *ibid.*, 1922, 120, 1744) represented the greatest novelty in a decade in which most new chloro- and bromo- derivatives of anthracene were prepared by classical methods.

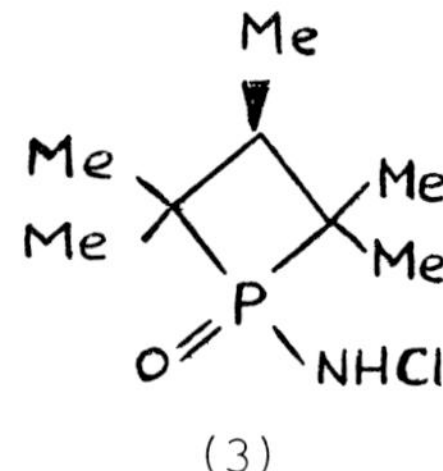

(3)

N-chloroamides such as (3) mimic sulphuryl chloride and chlorinate anthracene at C-9 (M.J.P. Harger and M.A. Stephens, J. chem. Soc., Perkin 1, 1980, 705). Copper(II) chloride (D.C. Nonhebel, J. chem. Soc., 1963, 1216) is a mild chlorination agent in solvents such as chlorobenzene and provides good yields of 9-chloroanthracene from anthracene. Its use, without solvents, has been advocated (I. Tanimoto, *et. al.*, Bull. chem. Soc. Japan, 1979, 52, 3586); under the latter conditions a quantitative yield of the 9-chloro product is reported. A mechanism is proposed for the reaction.

Iodination - The reaction of anthracene with iodine-$PhI(O_2CCF_3)_2$ did not result in iodination but in oxidation to give anthraquinone (E.B. Merkushev and N.D. Yudina, Zh. org. Khim., 19811, 17, 2598 (Chem. Abs. 1982, 96, 122304e). Suzuki's method (H. Suzuki, K. Nakamura, and R. Goto, Bull. chem. Soc. Japan, 1966, 39, 128) remains the most effective direct route to 9-iodoanthracene.

Cyanation of anthracene ($CNBr/AlCl_3$) gives an excellent yield (92%) of 9-cyanoanthracene (P.H. Gore, F.S. Kamounah, and A.Y. Miri, Tetrahedron, 1979, 35, 2927).

Bromomethylation (H_2CO/HBr) takes place at C-10 of a range of 9-substituted anthracenes; hexamethylenetetramine then provides the corresponding anthracenylmethylamine ($ArCH_2NH_2$) in good yield (N. Hartmann and M. Raethe, Z. Chem., 1979, 19, 373.

Nitration of anthracene to give 9-nitroanthracene may be brought about by NO_2 in dichloromethane (W.A. Pryor *et. al.*, J. org. Chem., 1984, 49, 5189). The parallel between relative rates and various parameters suggests the intermediacy of radical cations ($ArH^{+\cdot}$) rather than electrophilic attack upon arenes; anthracene, as in many other nitration studies,tended to give anthraquinone. Nitration of anthracene may also be encouraged by "graphite bisulphate" (J.P. Alazard, H.P. Kagan and R. Setton, Bull. Soc. chim. Fr., 1977 (5-6, Pt.2), 499). A mechanism of nitration of anthracene through electron-transfer has been proposed (L. Eberson and F. Radner, Acta. Chem. Scand., 1984, 38B, 861) in which the Wheland intermediate arises from the attack of the radical cation $ArH^{+\cdot}$ by NO_2; the ion itself has been prepared and characterised (esr) by photolysis of solutions of anthracene in trifluoroacetic acid (W. Chan *et. al.*, Chem. Comm., 1984, 1541).

Oxidation - Oxidation of anthracene generally leads to 9,10-anthraquinone. Thus iodosobenzene, like ozone, with anthracene gives the quinone, but a mixture of 9-formyl-10-methylanthracene (4) and 9-formylanthracene-10-carboxylic acid (5) with 9,10-dimethylanthracene. Ceric ammonium nitrate catalyses the two-phase oxidation (ammonium peroxydisulphate/sodium dodecylsulphonate/silver nitrate: J. Skarzewski, Tetrahedron 1984, 40, 4997). The presence of 18-crown-6 ether appears to promote oxidation by chromium(VI) oxide of a number of polybenzenoid hydrocarbons, including anthracene, to the corresponding quinone ((M. Juaristi *et. al.*, Can. J. Chem., 1984, 62, 2941). Correspondingly, the rhodium(II) acetate catalysed autoxidation of 1,4-dihydrobenzenes (molecular oxygen 1 atm., 25°) provides the arene, hydrogen peroxide, and water in many cases, but gives predominantly (75%) anthraquinone when applied to 9,10-dihydroanthracene

(M.P. Doyle *et. al.*, J. mol. Catalyst, 1984, 26, 259).

CH_3

CHO

(4)

CO_2H

CHO

(5)

Photochemical oxidation of anthracene derivatives, and the allied chemiluminescence process, have received intensive study. Two distinct mechanisms are reported for the photochemical oxidation of compounds such as 9,10-dicyanoanthracene in the presence of electron acceptors. One involves the production of singlet oxygen, while the other proceeds through the dioxygen radical anion ($O_2^{-\cdot}$) (J. Santamaria, Tetrahedron Letters, 1981, 22, 4511). The great range of chemiluminescent reactions makes their inclusion here selective. Compounds such as 1,5- and 1,8-bis(2-methoxycarbonylpropionyl)-anthracene, 9-cinnamoyl- and 9,10-dicinnamoyl-anthracenes, and 1,5-di-isobutyryl-anthracene are each reported to show chemiluminsence in their aerial oxidation in di-n-butyl phthalate containing potassium t-butoxide (Japanese Patent; see Chem. Abs., 1981, 94, 121175h; 121179q; 47023r). In the acyl group RCO- which is attached to both C-9 and C-10 of anthracene, R must be alkyl, aryl groups inhibit or reduce the chemiluminescent effect (I. Kamiya and T. Sugimoto, Photochem. Photobiol., 1979, 30, 49). The process was explained (T. Hiramatsu, T. Harada, and T. Yamagi, Bull. chem. Soc Japan, 1981, 54, 985) by a chemically initiated electron-exchange luminescence process (see also M.L. Vega, U.S.P., 4,076, 645; M.M. Rauhut, U.S.P. 4,064,064 (Chem. Abs. 1978, 88, 128926). Similar chemiluminescence occurs in the oxidation of a number of oxalate esters in the presence of anthracene derivatives (*e.g.* C.L.R. Catherall, T.F. Palmer, and R.B. Cundall, J. chem. Soc., Faraday 2, 1984, 80, 836), the colour of the emitted light being changed by the presence of dyes, usually derived from anthraquinone derivatives.

The photo-*reduction* of anthracene by sodium borohydride in ethanol, in the presence of a tertiary amine, gives

9,10-dihydroanthracene and 9,9',10,10'-tetrahydro-9,9'-bianthryl (N.C.C. Yang, W.L. Chang and J.R. Langan, Tetrahedron Letters, 1984, 25, 2855). The photo-Birch reduction ($NaBH_4$, $NaBH_3CN$) of anthracene has also been explored (M. Yasuda, C. Pac, and H. Sakurai, Chem. Abs., 1980, 93, 94535y). Cation radicals attack the borohydrides to give adducts whose structure depends upon the nucleophilicity and the steric requirements of the hydrides.

Hydrogenation of anthracene provides many polyhydro-derivatives whose stereochemistry arouse considerable interest; for example, the configurational analysis of 1,2,3,4,4a,9,9a,10-octahydroanthracene and its methyl derivatives (S. Hagishita and K. Kuriyama, Bull. chem. Soc. Japan, 1982 55, 3216). The chemistry of 1,2,3,4,5,6,7,8-octahydroanthracene was thought to be well known, but the nitro-derivative, formed by using methyl nitrate in the presence of boron trifluoride solutions in nitromethane, has m.p. 107-8° and not, as previously reported 89° (Sir John Cornforth, A.F. Sierakowski and T.W. Wallace, J. chem. Soc., Perkin 1, 1982, 2299). The identity of this nitro product as 9-nitrooctahydroanthracene was established by the presence of a signal due to one aromatic proton in its ^{1}H nmr spectrum. Conventional reactions converted this compound into the 9-iodo derivative, m.p. 67-8° (previously, m.p. 72-3°) which was also formed along with the 9,10-diiodo compound (m.p. 204-7°) by applying Suzuki's iodination method to the parent octahydroanthracene (H. Suzuki, K. Nakamura and R. Goto, Bull. chem. Soc. Japan, 1966. 39, 128).

Irradiating anthracene gives the photo-dimer, and this behaviour is shared by many of its derivatives, such as 9-methyl-, 9-chloro-, and 9-cyanoanthracene (J. Ferguson and S.H.E. Miller, Chem. Abs. 1976, 84, 126310q) although sandwich pairs are formed on irradiation short wave u.v. light. The range of peculiar structures which arise from irradiation of anthracene includes the lepidopteranes (H.D. Becker, K. Andersson and K. Sandros J. org. Chem., 1980, 45, 4549) and the 'jaws' structures (A. Castellan, J.M. Lacoste and H. Bouas-Laurent J, chem. Soc., Perkin 2, 1979, 411). Irradiation of anthracene in the presence of tetrafluorophthalonitrile in dichloromethane provides 9-(2,5,6-trifluoro-3,4-di-cyanophenyl)anthracene (K.A.K. Al-Fakhri and A.C. Pratt, Chem. Abs., 1977, 99, 212088d). Presumably hydrogen fluoride is lost from the initial addition product, since in methanol the product is 9-methoxy-10-(2,5,6,-trifluoro-3,4-dicyanophenyl)-9,10-dihydroanthracene (S. Yamada, Y. Kimura and M. Ohashi, Chem. Comm., 1977, 667). 1,2,4,5-Tetracyanobenzene reacts analogously, (Scheme 3) and the same substitution takes place in phenanthrene, also at C-9 (S. Yamada and M. Ohashi, Chem. Abs., 1984, 100, 191531q).

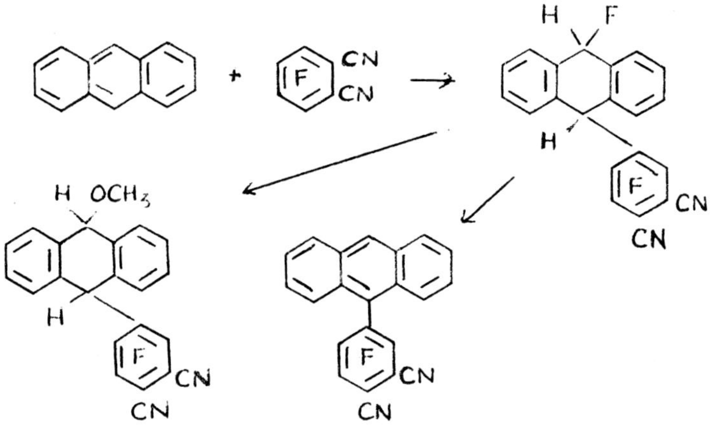

Scheme 3

(c) Meso-addition

Addition across the *meso*-positions of anthracene is probably its most characteristic reaction. Photo-catalysed *meso*-addition of acrylonitrile to anthracene shows a higher quantum yield in polar solvents, which suggests a rate-determining proton-transfer to the exciplex (N. Selvarajan and V. Ramakrishnan, Indian J. Chem., 1979, 18B, 331). The *meso*-addition of *N*-phenylmaleimide to anthracene is catalysed by gallium trichloride, allegedly because of the lowering of the LUMO energy of the dienophile upon nV-complexation (V.D. Kiselev, I.M. Shakirov, and A.I. Komovalov, Zh. org. Khim., 1984, 20, 1454). The reactions with quinones were found to show a greater susceptibility to such catalysis (V.D. Kiselev, D.E. Khuzyasheva and A. I. Konovalov, Zh. org. Khim., 1983, 19, 1268), and the influence of π-donor solvents upon the process was also reported (D.G. Khuzyasheva, *et. al.*, *ibid.*, 1983, 19, 1617). The addition of trichlorogermane to 9-methylanthracene (benzene, oxygen) gives 9-methyl-9,10-bis-(trichlorogermanyl)-9,10-dihydroanthracene (S. Kolesnikov, I.V. Lyudkovskaya and O.M. Nefedov, Izv. Akad. Nauk SSSR, Ser. Khim., 1983. 6, 145) though 9-10-dihydro-9-methyl-9-(trichlorogermanyl)anthracene arises in the absence of oxygen (S.P. Kolesnikov, L.V. Lyudkovskaya and O.M. Nefedov, *ibid.*, 1983, 7, 1612). The Diels-Alder adduct of tetranitrobutadiene and anthracene arises from the treatment of a mixture of anthracene and 1,1,4,4-tetranitro-2,3-diacetoxybutane with base (G.V. Nekrasova *et. al.*, Zh. org. Khim., 1984, 20, 2502). The addition of vinylene carbonate to anthracene provides *cis*-9,10-ethanoanthracene-11,12-diol cyclic carbonate which provides 9,10-dihydro-9,10-diformylanthracene upon hydrolysis followed by oxidation with lead tetraacetate (R.G. Child, Chem. Abs., 1982, 96, 217841s). This route to 9,10-diformylanthracene affords a new source of the precursor to 11,11,12,12-tetracyanoanthraquinone dimethide, the TCNQ analogue, and has been assessed alongside others (K.C. Murdock, *et. al.*, J. med. Chem., 1982, 25, 505). Particular structures, such as the [4,4,2]propellanes, are also available by Diels-Alder processes (M. Oda and Y. Kanao, Chem. Letters, 1981, 1547). The sensitivity of the rate of addition of tetracyanoethylene to substituted anthracenes depends upon the position of the substituent. At C-10 and C-9- susbstituents show $\rho = -7$, but substituents at C-1 or C-2 show only $\rho = -3$ (R.M.G. Roberts and F. Yavari, Tetrahedron, 1981, 37, 2657). The difference arises from the different facilities with which substituent effects may be transmitted to the reaction site. Perturbation of two aromatic rings is a higher energy process and is more resisted; a similar effect has been seen in the chlorination of anthracene derivatives by sulphuryl chloride.

Anthracenes and their derivatives may be made by Diels-Alder processes. The addition of dihalogenocyclopropene (6) across the diene system of 2,3-bis(methylene)-1,2,3,4-tetrahydronaphthalene provides a route to 1H-cycloprop[b]anthracene (7) (W.E. Billups, E.W. Casserley and B.E. Arney, jr., J. Amer. chem. Soc., 1984, 106, 440).

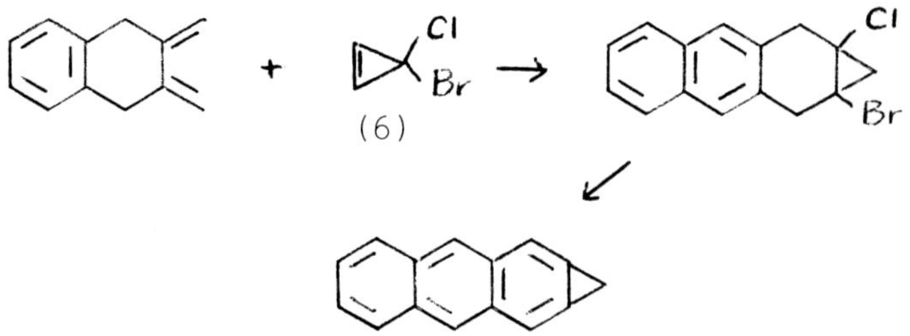

Cl Br (6) Cl Br (7)

Scheme 4

Triptycenes from anthracene derivatives. - The stereochemical interest in 'gearing' of different substituents, free to rotate but physically interacting with substituents, has encouraged the synthesis of a range of triptycenes, almost always by the addition of arynes across the *meso*-position of substituted anthracenes. Both positive and negative buttressing have been found in the species formed from the addition of tetrahalogenobenzynes to 9-(1,1-dimethyl-2-phenylethyl)anthracene (G. Yamamoto, M. Suzuki and M. Oki, Bull. chem. Soc. Japan, 1983, 56, 306 and 809). A number of hindered 9-benzylanthracenes, leading to a set of hindered triptycenes, have been used to achieve conformational mapping (R.B. Nachbar, jr., *et. al.*, J. org. Chem., 1983, 48, 1227; M. Oki, *et. al.*, Bull. chem. Soc. Japan, 1983, 56, 302.

2. *Substituted anthracenes*

(a) *Halogenoanthracenes*

The homolytic phenylation of 9-chloro- and 9-methylanthracene by benzoyl peroxide has been reported (F.M. Cromarty, R. Hendriquez and D.C. Nonhebel, J. chem. Res., (S), 1977, 309), and the photodimerisation of 9-chloroanthracene has already been mentioned (*v.i*). 9-Bromo-10-X-anthracene derivatives react with potassium phenoxide in dimethylformamide to give the expected S_NAr products if X is activating (*e.g.* X = CN) but much reduction if not (*e.g.* X = Ph). 9-Bromoanthracene itself gives (NaOMe-pyridine) 45% anthracene, and the use of sodium trideuteriomethoxide causes C-9 to become specifically labelled. (J. Rigaudy, A.M. Seuleiman, and K.C. Nguyen, Tetrahedron, 1982, 38, 3157). The same authors report (*idem. ibid.*, p.3151) that 9-bromo-10-ethyl- or -10-benzylanthracene react with potassium phenoxide in HMPT to give mixtures in which the phenoxy group also appears at the α-position in the side-chain, a phenomenon which they call 'tele-substitution'.

(b) *Acylanthracenes*

9-Formylanthracene reacts with ethanolic potassium cyanide to give 9-cyanoanthracene. 9-Cyano-10-formylanthracene is the suggested intermediate, and presumably decarbonylation occurs *via* a carbanion (P.H. Gore, S.D. Gupta and G.A. Obaji, J. prakt. Chem., 1984, 326, 381). The synthesis of 9,10-diformylanthracene, a precursor of anthraquinonedimethide systems, has been reviewed (K.C. Murdock, *et. al.*, J. med. Chem., 1982, 25, 505.

(c) Anthraquinones

(i) Applications. Anthraquinone derivatives are widely used as dye components; such dyes have been incorporated into liquid-crystal display systems, as in the trans-4-butyl-cyclohexyl esters of 1-amino-4-nitro-anthraquinone-2-carboxylic acid (M. Kaneko, Chem. Abs., 1983, 99, 46062u); they also have application in photothermography (Ricoh Co., Jap. Pat., Chem. Abs., 1983, 99, 13989 and 13993). Among more imaginative uses of these derivatives is the use (R.A. Kenley, Chem. Abs., 1983, 99, 157401a) of 1-amino-4-hydroxyanthraquinone as a photosensitizer, which, in a polymer, stimulated photo-oxidation of toxic warfare gases. Mustards were coped with well, but nerve gases were only poorly deactivated. Derivatives of 5,8-dihydroxy-1,4-bis(2-alkylethylamino)anthraquinone inhibit the growth of neoplasts (K.C. Murdock, Chem. Abs., 1983, 99, 194946a).

9,10-Anthraquinones are stable species, and are readily formed in a range of oxidative conditions. Thus 2,5-dialkyldiphenylmethanes give 2-alkylanthraquinones when passed over V_2O_5/Ce_2O_3 on alumina at 500° (Pat., Chem. Abs., 1977, 94, 121174j). Benzoylbenzoic acid derivatives are convenient sources of 2-alkylanthraquinones (S.B. Bowlus, Synth Comm., 1984, 14, 391).

Derivatives of anthraquinone made by Diels-Alder reactions have been reviewed (J.A. Lowe, *et. al.*, Tetrahedron, 1984, 40, 4751), as have the synthetic methods available for anthraquinone-type natural products (V. Guay and P. Brassard, Tetrahedron, 1984, 40, 5039).

The synthesis of 11,11,12,12-tetracyanoanthraquinone dimethide (8) has been achieved by conventional processes (Scheme 5) Pat., Chem. Abs., 1983, 98, 197809y) and by the obvious route of condensing anthraquinone with malononitrile ($TiCl_4$/pyridine) (B.S. Ong and B. Keoshkerian, J. org. Chem., 1984, 49, 5002); the most recent of the three nearly coincident reports (A.M. Kini, J. Amer. chem. Soc., 1985, 107, 556) relies on another well-established synthetic process involving 9,10-bis-(cyanomethyl)anthracene as a key intermediate (Scheme 6). The synthesis of a number of derivatives of both 1,4- and 9,10-anthraquinone containing $=C(CN)_2$ fragments has been reported, together with the properties of their tetrathiafulvalene and tetrathiatetracene charge transfer complexes (Y. Sakata, et. al., Chem. Abs., 1984, 101, 170414j).

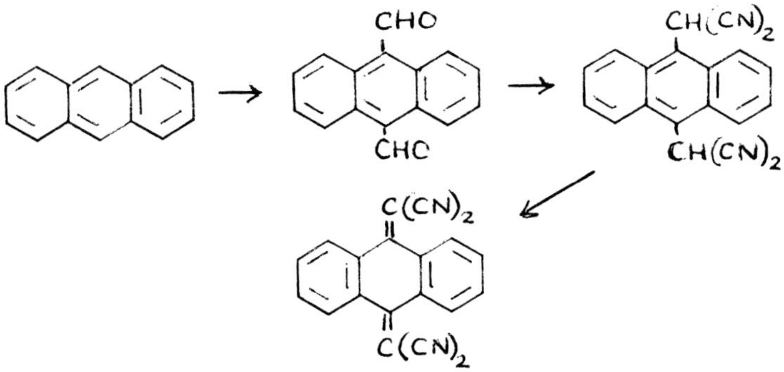

(8)

Scheme 5

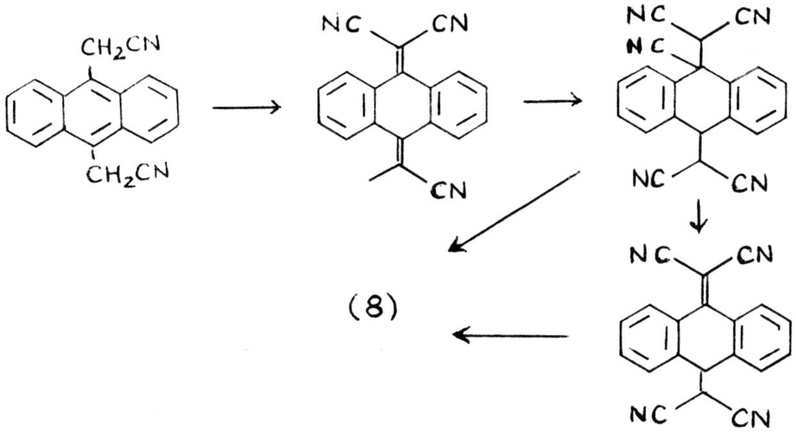

(8)

Scheme 6

The ease of exchange of functionalisation between groups in 1,4-dihydroxyanthraquinone, and in some of its derivatives, is exemplified by the migration of the acetyl group in 1-acetoxy-4-amino-8-hydroxyanthraquinone(9), under the influence of base, to give the 1-hydroxy-8-acetoxy isomer (10) (V.P. Volosenko and S.I. Popov, Zh. org. Khim., 1981, 17, 874). Similarly, both 2-methyl- and 2-chloro-1,4-dihydroxy-9,10-anthraquinone react with thionyl chloride to give a mixture of the 2- and the 3-substituted-9-hydroxy-10-chloro-1,4-anthraquinones (11) which then provide the corresponding 1-hydroxy-4-chloro-9,10-anthraquinones (12) (Scheme 7) (M.V. Gordik, *et. al.*, Zh. org. Khim, 1984, 20, 1934).

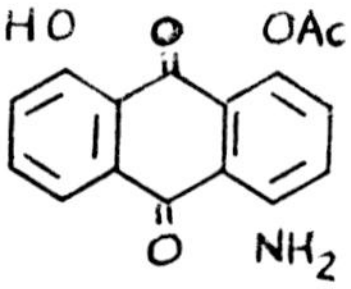

HO O OAc

O NH_2

(9)

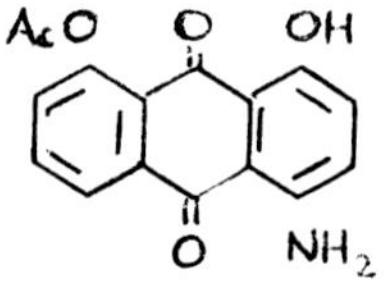

AcO O OH

O NH_2

(10)

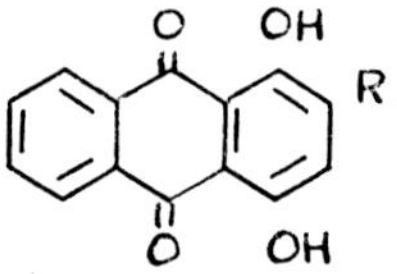

O OH

R

O OH

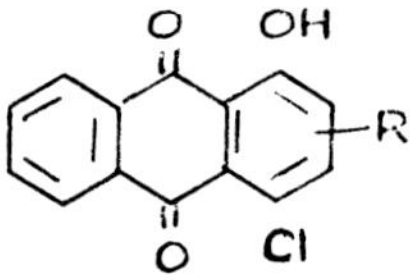

O OH

R

O Cl

(12)

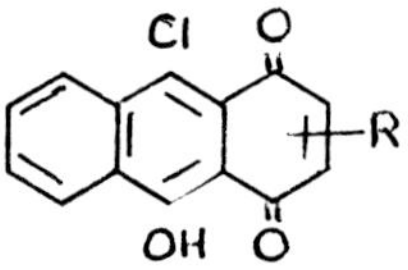

Cl O

R

OH O

(11)

Scheme 7

The effects of substitutents upon ozonolysis of anthraquinones has been reviewed (M. Matsui, Chem. Abs., 1984, 101, 2319261).

(ii) Substitution

The carbonyl functions in anthraquinone activate the system towards nucleophiles, so that many of the great range of recently reported anthraquinone derivatives come from such processes. Thus, 1,4,5,8-tetrachloroanthraquinone gives yellow or red dyes by displacement of halogen with thiophenoxide ions (R. Neeff, Chem. Abs., 1983, 99, 89624w). The treatment of 1-amino-4-phenylamino-anthraquinone-2-sulphonic acid with cyanide ion in aqueous solution provides a mixture of the 2-cyano derivative and the desulphonated derivative, but in dimethyl sulphoxide the 2,3-dinitrile is formed. The nitriles are most susceptible to attack by nucleophiles such as aliphatic amines, CN being lost as the anion (J.M. Adam, and T. Winkler, Helv., 1983, 66, 411). 1,4-Diaminoanthraquinone gives the diamino-2,3-dicyano derivative with alkali cyanides (J.M. Adam, Chem. Abs., 1983, 99, 124081r) which presumably involves nucleophilic addition and then oxidation of the original adduct.

Polychloro-anthraquinones upon irradiation of their ethanolic solutions at wavelengths of 365 nm lose chlorine specifically from the α-position; β-chlorine is not displaced (H. Inoue, K. Ikeda and M. Hido, Chem. Abs., 1984, 100, 191536v).

Iodination of aminoanthraquinones is brought about by I_2/HIO_3 in acetic-sulphuric acid mixture at 70°. 1-Amino-anthraquinone gives 2-iodo- and 2,4-di-iodo-1-amino-anthraquinone while 2-amino-anthraquinone gives 1,3-di-iodo-2-amino-anthraquinone. 2-Amino-3-chloro-anthraquinone gives 2-amino-3-chloro-1-iodo-anthraquinone (A.A. Moroz and I.A. Beloborodova, Zh. org. Khim., 1981, 17, 2612).

Interphase catalysts improve the yield of 1-phenyl-anthraquinone from anthraquinone-1-diazonium borofluoride from 7% in their absence to 80-90% in their presence; all the phase-transfer agents used had similar efficacies (L.L. Puchkina, *et. al.*, Zh, org. Khim., 1983, 19, 1486).

(d) Phenanthrene

(i) Occurrence

Phenanthrene has been found, for example, in snow and rain in Norway (J. Peeregaard and E.J. Schiener, Chem. Geol., 1979, 26, 331); in the North Atlantic (A. Saliot, M.J. Tissier and C. Boussuge, Phys. Chem. Earth, 1980, 12, 333); in Switzerland and in Lake Washington, USA, (S.G. Wakeham, C. Schaffner and W. Giger, Phys. Chem. Earth., 1980, 12, 353); in Japanese river waters (R. Ishiwateri and T. Honya, Chem. Abs., 1977, 86, 60292v) and its global incidence has been reported (R.E. Laflamme and R.A. Hites, Chem. Abs., 1979, 89, 114246c).

(ii) Structure

The close proximity of substituents at C-4 and C-5 leads to crowding and their anomalous behaviour in a variety of conditions. The ^{1}H-nmr spectrum of 4-ethynylphenanthrene (13) shows an unusually low-field proton signal associated with C-5 (F.B. Mallory and M.B. Baker, J. org. Chem., 1984, 49, 1323), whereas 4-methylphenanthrene has been reported (K. Takegoshi, *et. al.*, J. chem. Phys., 1984, 80, 1089) to show a high (5.05 kcal mole^{-1}) barrier to rotation about the C-4/CH_3 bond. Similarly, the synthesis of 4,5-difluorophenanthrenes has allowed the nmr investigation of the crowding in this system (F.B. Mallory, C.W. Mallory, and W.M. Ricker, J. org. Chem., 1984, 50, 457), extending the earlier work on 1-X-8-methyl-4,5-difluoro phenanthrenes (K.L. Servis annd K.-N. Fang, J. Amer. chem. Soc., 1968, 90, 6712).

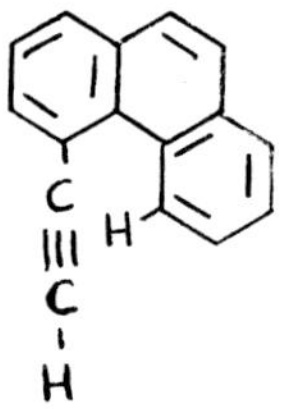

(13)

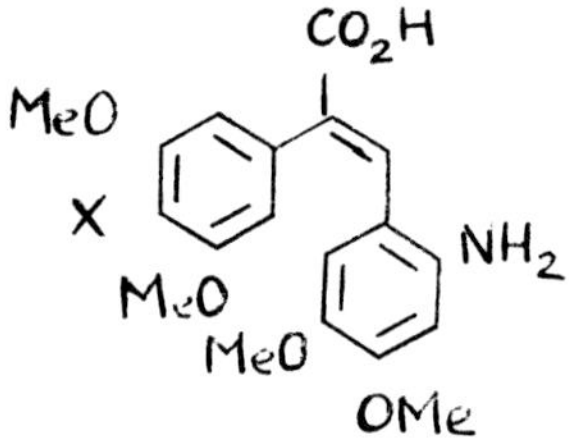

(14)

(iii) Synthesis

The classic Pschorr reaction (R. Pschorr, Ber., 1896, 29, 496) has been refined by a number of amendments. In the presence of sodium iodide, which has been shown (B. Chauncy and E. Gellert, Austral. J. Chem., 1969, 22, 993) to encourage the homolytic ring closure ..., the yield of cyclised product from diazotised 2-amino-3',4,5,5'-tetramethoxy-4'-X-phenylcinnamic acid (14) is improved when X is Ph.SO_2.O- (14a) and not when X is MeO (14b) (R.I. Duclose, jr., J.S. Tung, and H. Rapoport, J. org. Chem., 1984, 49, 5243). The photo-catalysed Pschorr reaction has been reported to give quantitative yields, (H. Cano-Yelo and A. Deronzier, J. chem. Soc., Perkin 2, 1984, 1093) with high quantum yield (> 0.4) in the presence of Ru(2,2'-bipyridyl)$_3^{2+}$.

Less widely useful synthesis of phenanthrene include the pyrolysis of bis-(butenynyl)benzene (B.C. Berris and K.P.C. Vollhardt, Tetrahedron, 1982, 38, 2911), when the formation of phenanthrene is accompanied by a range of hydrocarbon products. Pyrolysis of 9,10-dihydro-9-(N,N-dimethylamino)phenanthrene, from benzyne and N,N-dimethyl-aminostyrene provides phenanthrene and dimethylamine (L.N. Koikov, P.B. Terent'ev and Yu. G. Bundel, Zh. org. Khim., 1983,, 19, 1552).

The synthesis of a mixture of 1,4- and 9,10-dihydro-phenanthrenes by the light-catalysed cyclisation of 1,2-diarylethenes in amines, involves a base-catalysed proton jump from C-4b to C-4a in the 4a,4b-dihydro-phenanthrene which is the directly-formed product of photolysis. In base/methanol solutions, the selectivity of formation is less. With thiopropoxide as base, deprotonation-protonation only occurs at C-9, so that 9,10-dihydrophenanthrene is produced, but again with the stronger base MeO^- the selectivity is less. (J.B.M. Somers *et. al.*, J. Amer. chem. Soc., 1985, 107, 1387). The mechanism of these photo-catalysed cyclisations has been suggested to involve a radical cation (L.W. Reichel *et. al.*, Can. J. Chem., 1984, 62, 424). Photo- chemical synthesis of polybenzenoid aromatic hydrocarbons, including phenanthrene, has been discussed critically (J. Brison *et. al*, Bull. Soc. chim. Belg., 1983, 92, 901), and the cleavage of nitro substituents from 1,2-dinitro-1,2-diphenylethane under irradiation from a high-pressure mercury lamps offers a route to stilbene and hence to phenanthrene (K. Fukunaga and M. Kimura, Chem. Abs., 1984, 100, 191531q). This well-known process was extended by the synthesis and irradiation of 2-styryl-9,10-dihydro-phenanthrene (15) which gives the dihydro derivatives of dibenzoanthracene (16) and benzochrysene (17) (F. Diederich, K. Schneider and H.A. Staab, Ber., 1984 117, 1255), though *cis*-1,2-(9-anthracenyl)ethene gives only the anthracene photo-dimers (H.D. Becker, K. Sandros, and K. Andersson, Angew. Chem., 1983, 95, 507).

1,8-Dimethylphenanthrene may be made from *o*-iodo-toluene through 1,2-bis(*o*-methylphenyl)ethyne (18) ($C_2H_2/PdCl_2/PPh_3$ in boiling Et_2NH followed by selective hydrogenation and photolysis in the presence of iodine

(27% overall: T.S. Skorokhodoova and V.I. Karpitskii, Chem. Abs., 1984, 100, 156333e).

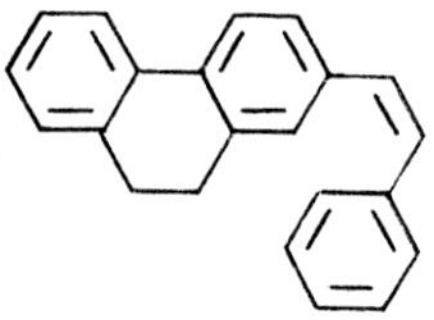

(15)

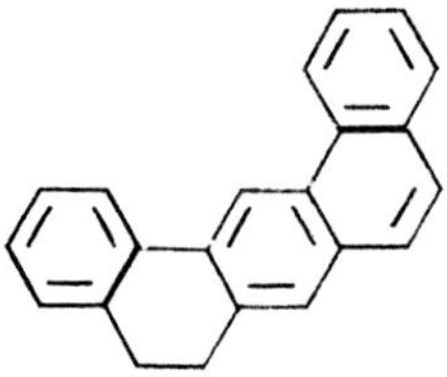

(16)

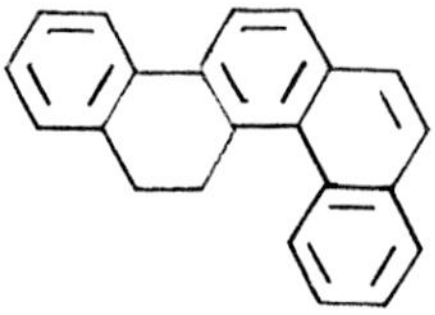

(17)

Me Me

(18)

A variant on this process is the treatment of Z-2-chlorostilbenes with activated magnesium ($MgCl_2$/K/THF; 12h. reflux) to form phenanthrene. A 2'-methoxy group, which appears at C-1 of the product, does not inhibit the process (76-83% yield) (C. Brown, *et. al.*, J. chem. Soc., Perkin 1, 1982, 3007). The formation of 9,10-dihydrophenanthrene from 2,2'-bis(chloromethyl)biphenyl may be achieved using magnesium or lithium in tetrahydrofuran when various complexes seem to be involved (L.M. Engelhardt, W.-P. Leung, C.R. Raston, P. Twiss and A.H. White, J. chem. Soc., Dalton, 1984, 331).

Phenanthrene formation may also be achieved from (cyclohexen-1-yl)cyclohexanone which with perpropionic acid gives the furan derivative (19). This undergoes a Diels-Alder reaction with, for example, acrylic acid to give a product which on heating with palladised charcoal, provides phenanthrene (O. Weiberg, H. Seifert and H. Waldmann, Chem. Abs., 1983, 99, 5503v)

(iv) Substitution

Ozonolysis of polycyclic hydrocarbons, including phenanthrene, is reported (W.A. Pryor, G.J. Glericher and D.F. Church, J. org. Chem., 1983, 48, 4198) to involve a slow electrophilic attack, as judged by the change of selectivity with solvent; the rate-determining formation of pi or sigma-complexes was also suggested, although the two processes may well have similar electronic demands. Ozonolysis of phenenathrene in cold chloroform-methanol provides (20) in 84% yield. The latter with *n*-butylamine provides a Schiff's base which, with nitromethane, gives methyl 2'(2-nitroethenyl)biphenyl-2-carboxylate (F. Szemes and A. Rybar, Chem. Abs., 1984, 100, 68192c).

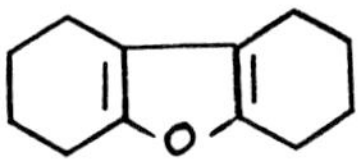

(19)

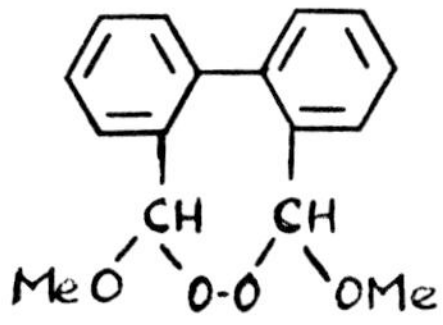

CH CH
MeO O-O OMe

(20)

R'
R
O
O

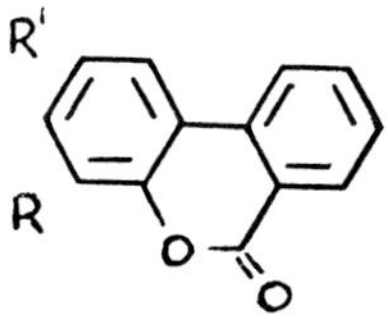

(21)

However, ozonolysis of phenanthrene in (HCl-MeOH) gives the dibenzopyranones (21) (R,R' may be H or Cl) as well as dimethyl diphenate (J. Neumeister and K. Griesbaum, Ber., 1984, 117, 1640.

In an environmentally significant study, the high rates constant for ozonolysis of phenanthrene in aqueous media (k_2, ca. 10^4 dm^3 mol^{-1} s^{-1}) suggested a half-life of less than a second (V. Butkovic *et. al.*, Chem. Abs., 1983, 99,

87459x) but despite this deduction the incidence of phenanthrene world-wide, though not ubiquitous, is general.

Nitration of phenanthrene by ceric ammonium nitrate in acetic acid is advanced (H.M. Chawla and R.S. Mittel, Indian J. Chem., 1983, 22B, 753) as a one-pot route to 2-nitrophenanthrene.

Radiation-induced hydroxylation of phenanthrene, among other polycyclic aromatic hydrocarbons, proceeds by a slow formation of the radical cation by electron transfer (L.N. Bortun, A.F. Rekasheva, Chem. Abs. 1983, 99, 157513m).

Photo-cyanation of phenanthrene involves two photo-induced transients ((H.J. Lemmetyinen, *et. al.,* J. Photochem., 1983, 22, 115; H.J. Lemmetyinen, J. chem. Soc., Perkin 2, 1983, 1269. An ionic complex may arise through a triplet excimer of ArH or, in the presence of an electron-acceptor, a triplet exciplex, and this transient complex takes up CN- to give $[ArHCN]^{\cdot}$. In aqueous acetonitrile this disproportionates, forming also the dihydronitrile, but in dry acetonitrile oxidation occurs.

Bromine azide adds as expected across the 9,10-bond of phenanthrene, but the product upon treatment with $LiAlH_4$ is 9,10-iminophenanthrene; this reaction sequence can afford a convenient route to 9-aminophenanthrene (85%) (J.W. Denis and A. Krief, Tetrahedron, 1979, 35, 2901). 9,10-Iminophenanthrene reacts with alkyl halides in the presence of sodium azide or ammonium thiocyanate; depending upon the nature of the quaternary ammonium ion also present, yielding 9-alkyliminophenanthrenes (22) or 9-amino-10-azido-9,10-dihydrophenanthrene, whereas KSCN provides (23) or, with the N-butyliminophenanthrene, (24) (M. Weitzberg, Z. Aizenshtat, and J. Blum, J. heterocyclic Chem., 1983, 20, 1019).

9-Chloroimino-phenanthrene gives 10,10-dialkoxy-9-(10H)-phenanthrone with silver ion in alcohols (S. Shtelzer, *et. al., ibid.,* 1984, 21, 1593).

Cycloaddition occurs with phenanthrene, although less generally than with anthracene; the addition of 1,1-, 1,2-, and 1,1,2-polychloroethanes may be photolytically induced to give [2+2]-adducts which aromatise (NBS) to give cyclobuta[l]phenanthrene (25) (N.P. Hacker and J.F.W. McOmie, Tetrahedron, 1984, 40, 5249).

The synthesis of methyl phenanthrene-9-carboxylate from 9-(selenophenyl)phenanthrene upon treatment with dichlorocarbene is held to involve the intermediacy of the three-membered intermediate (26) (Scheme 8) (B.R. Dent and

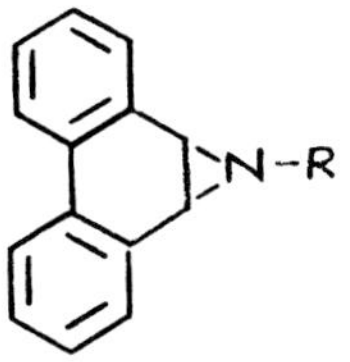

(22) 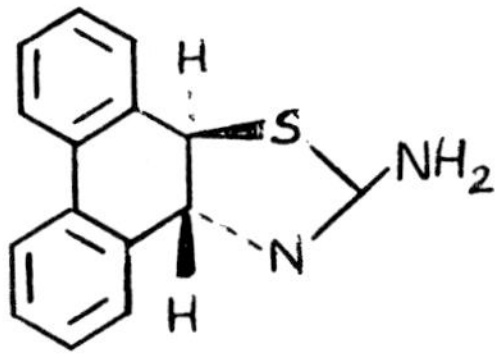 (23)

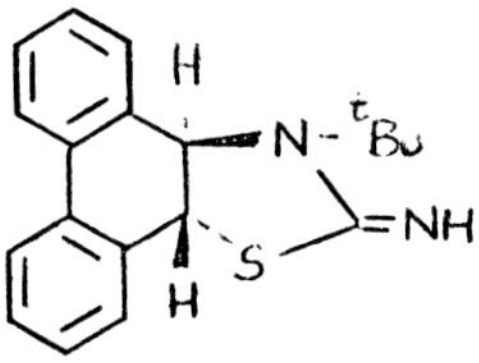

(24) 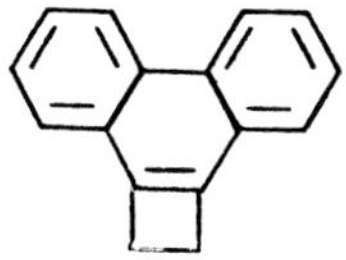 (25)

B. Halton, Tetrahedron Letters, 1984, 25, 4279). In addition of :$CH(CO_2Et)$ to phenanthrene, Rh(III)-porphyrins appreciably assist the formation of the adduct (H.J. Callot, F. Merz and C. Piechicki, Tetrahedron, 1982, 38, 2365). The synthesis of methyl phenanthrene-3-carboxylate from the product of Friedel-Crafts acylation (oxalyl chloride) of phenanthrene (23%) has been reported (D.W. Miller, *et. al.*, J. chem. Res. (S), 1984, 418). Other annelation processes include the synthesis of phenanthro[9,10,-b]- thiophen (27) through a ring-closure of 9-phenanthrylpropenoic acid (28) by thionyl chloride. The compound (28) was obtained from 9-bromophenanthrene and acrylic acid (T.N. Sidorenko, G.A. Terent'ova and V.S. Aksenov, Chem. Abs, 1983, 99, 5463g).

Phenanthrene-9-carboxyaldehydes are convenient precursors of a number of natural products, and the synthesis of 2,3,6-trimethoxyphenanthrene-9-carboxyaldehyde leads to a synthesis of cryptopleurine (M.L. Bremmer, N.A. Khatri and S.M. Weinreb, J. org. Chem., 1983, 48, 3661; see also T.F. Buckley III, and H. Rapoport, *ibid.*, 1983, 48, 4222).

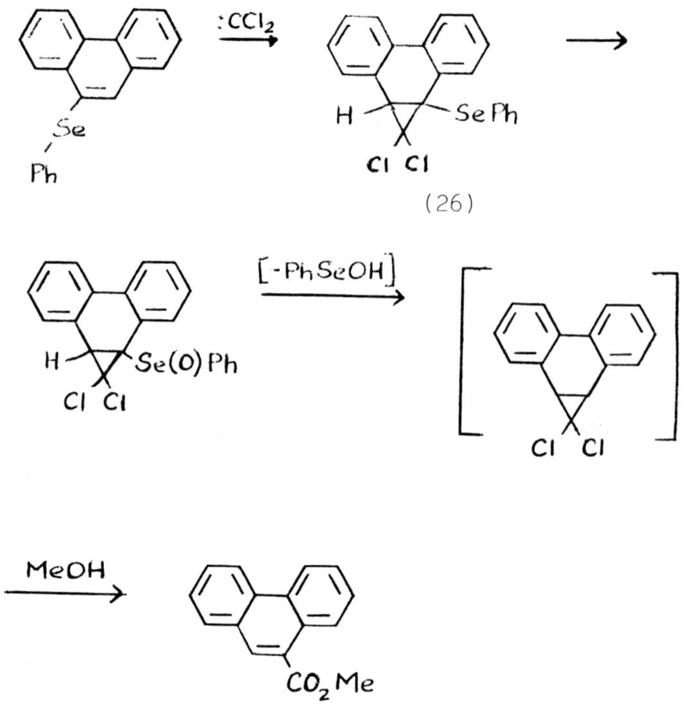

$:CCl_2$

(26)

[-PhSeOH]

MeOH

Scheme 8

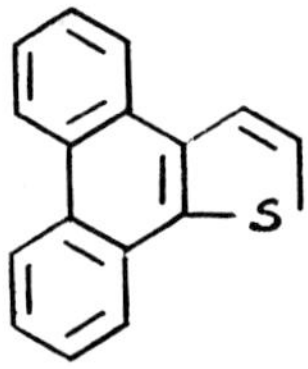

(27)

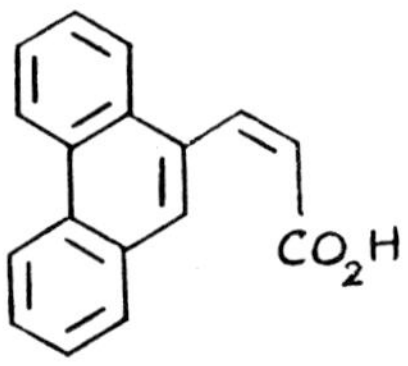

(28)

9,10-Phenanthraquinone is the source of some polyannular benzenoid systems through a bis-Wittig reaction with $C_6H_4(CH_2PPh_3)_2$ using phase-transfer catalysts to promote the reaction (A. Minsky and M. Rabinovitz, Synthesis, 1983, 6, 497). The extrusion of carbon monoxide from phenanthraquinone gives fluorenone. The thermolysis of phenanthraquinone (and anthraquinone) gives fluorenone and, in an atmosphere of hydrogen, biphenyl and benzene which is the major product, the kinetics of the process have been reported (T. Sakai, M. Hattori, and N. Yamane, Chem. Abs., 1981, 94, 138966a). The oxidation of phenanthraquinone by alkaline hydrogen peroxide has also been used to obtain fluorenone, and the process has been shown to involve the addition of the hydroperoxide anion to the carbonyl system and not the formation of a dioxetane or an epoxide intermediate. (Y. Sawaki and C.S. Foote, J. Amer. chem. Soc., 1983, 105 , 5035). Oxygenation of pyridine solutions of phenanthraquinone derivatives in the presence of coppper(I) chloride gives good yields of the corresponding diphenic acid; the corresponding reactions of phenanthraquinone monoxime gave 2-cyanobiphenyl-2'-carboxylic acid (E. Balogh-Hergovich, G. Speier and Z. Tyeklar, Synthesis, 1982, 731).

The Schmidt reaction (NaN_3-H_2SO_4) of phenanthraquinone is well reported, but in the presence of nitric acid at ca. 100° the process is a synthetic route to 4,4',6,6'-tetranitrodiphenic acid. At high temperatures (150°) nitrodecarboxylation gives rise to 2,4,4',6,6'-penta-nitrobiphenyl-2'-carboxylic acid (A.M. Andrierskii, A.N. Poplavskii and K.M. Dyumaev, Chem. Abs., 1979, 91, P.123519n; 1980, 93, P.150004p).

Iodination of phenanthraquinone (I_2-$KMnO_4$-H_2SO_4) gives 2,7-di-iodophenanthraquinone (V.K. Chaikovskii, A.N. Novikov and N.M. Dubovitskaya, Chem. Abs., 1982, 97, 162475p); an earlier report of the formation of 2-iodophenanthraquinone by iodination gives no details (A.N. Novikov, V.T. Slyusarchik and E.F. Matantseva, Chem. Abs, 1979, 89, 24025a).

Phenanthraquinone undergoes a variety of cycloadditions, such as that seen with isobenzofuran (W. Friedrichsen, I. Kallwert and R. Schmidt, Ann. 1977, 116).
Photochemically induced reactions of phenanthraquinones and their application to the preparation of photo-sensitive materials has characterised much of the recently published work. In photochemically induced radical oxidation

processes the observation of CIDNP has been found to be dependent upon the strength of the magnetic field; the phenanthraquinone - xanthone system was the example studied.

Chapter 29

FUSED-RING POLYCYCLIC AROMATIC COMPOUNDS CONTAINING ONE OR MORE FIVE-MEMBERED RINGS

H.F. ANDREW

The organisation of this Chapter is according to the number of "free" positions in the five-membered rings and follows that adopted in the Second Edition. The prototype hydrocarbons are benzindan, acenaphthene, fluorene and fluoranthene. A similar format is followed for compounds containing two five-membered rings, and finally aromatic hydrocarbons containing five- and seven-membered rings are considered.

1 The benzindan group

7H-Cyclopenta[a]pyrene (1) m.p. 125-127° and 9H-Cyclopenta[a]pyrene (2) m.p. 133-135° are both synthesised from pyrene-1-carboxaldehyde *via* the key intermediate (3) (H.Lee and R.G.Harvey, J.org.Chem., 1982, 47, 4364).

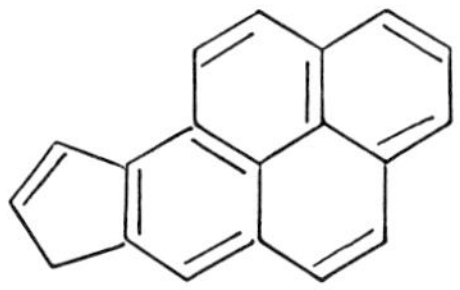

(1) 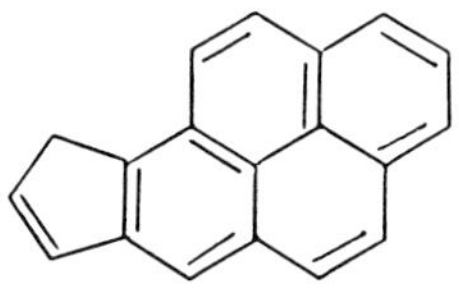 (2) 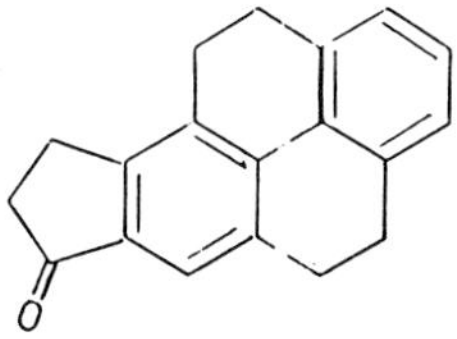 (3)

Wolff-Kishner reduction of (3) followed by aromatisation affords dihydrocyclopenta[a]pyrene, which is converted to (1) by successive treatment with DDQ, sodium borohydride and *p*-toluenesulphonic acid.

Reduction of (3) with sodium borohydride, followed by acetylation of the resulting alcohol and dehydrogenation furnishes an acetate which undergoes elimination to form (2). Hydrocarbon (2) readily forms an epoxide, but (1) does not.

Cyclopenta[a]phenalene (4), a non-alternant hydrocarbon of considerable theoretical interest, is as yet unknown but the simple derivative (5) has been prepared. Synthesis of (5) is accomplished in eight steps from phenalene. A key step in the synthesis is ring-expansion of the cyclobutanone (6) by means of diazomethane to give (7). (Y.Sugihara, H. Fujita and I.Murata, Chem.Comm. 1986, 1130).

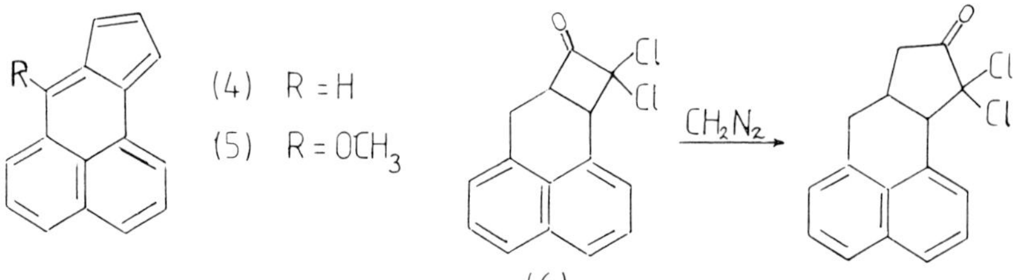

(4) R = H

(5) R = OCH_3

CH_2N_2

(6)

(7)

The hydrocarbon (5), dark brown needles m.p. 140-141°, has azulene-like character in the ground state. Its structure is supported by spectroscopic evidence, a noteworthy feature of its proton nmr spectrum being the appearance of the five-membered ring protons at lower field than those of fulvenes suggesting that (5) is semi-aromatic.

2 The acenaphthene group

(a) Acenaphthene and derivatives

The interesting regioselectivity of acyl fluorides in the presence of BF_3 is the basis of a useful route to a variety of 3-substituted acenaphthenes (J.A.Hyatt and P.W. Reynolds, J.org.Chem., 1984, 49, 384). Thus acenaphthene with $(CH_3)_2CHCOCl$ and $AlCl_3$ gives the 5- and the 3- ketone in the ratio 80:20, whereas $(CH_3)_2CHCOF$ and BF_3 give the same products in the ratio 15:85.

Acenaphthene dithione (8) is an aromatic α-dithione with rigid cisoid, coplanar C = S groups; and the ring-tautomeric dithiete should be disfavoured (M.Cava *et al.*, J.org.Chem., 1985, 50, 1550). Photolysis of (9) however, does not produce (8), but gives instead the 1,4-dithiine (10).

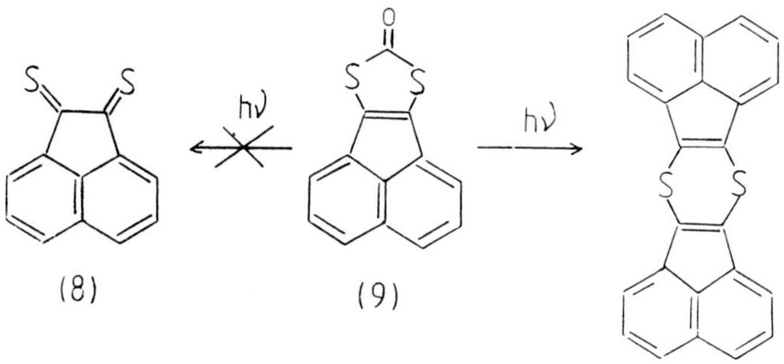

(8) (9) (10)

Trapping experiments with, for example, norbornene, establish that (8) is in fact formed as a strained intermediate.

Tetrahydro-2a,3,4,5-acenaphthene (11) provides a useful vehicle for studying effects of ring-strain and hyperconjugation on the electrophilic substitution reactions of the benzene ring (R.Gruber, D.Cagniant and P.Cagniant, Bull.Soc.chim.Fr., 1977, 773).

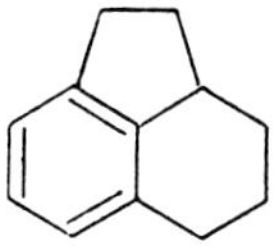

(11)

Paracyclophanes (12,13) containing the acenaphthene nucleus are synthesised by intramolecular reductive coupling of corresponding di-carbonyl compounds, using low-valent titanium reagents (A.Kasahara *et al.*, Bull.chem.Soc.Japan, 1982, 55, 2434).

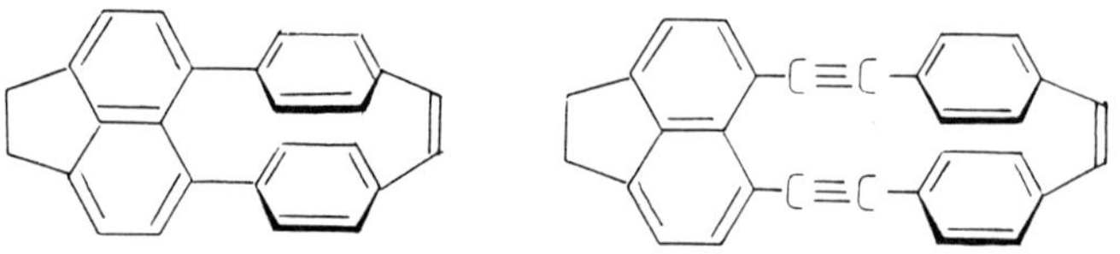

(12) (13)

Such compounds reveal, through their electronic spectra, useful information on transannular π-electronic interactions.

(*b*) *Acenaphthylene and derivatives*

Acenaphthylene (16) is formed when the diazoacenaphthene (14) decomposes thermally, presumably *via* the carbenic intermediate (15).

(S-J.Chang, B.K Ravi Shankar and H.Shechter, J.org.Chem., 1982, 47, 4226)

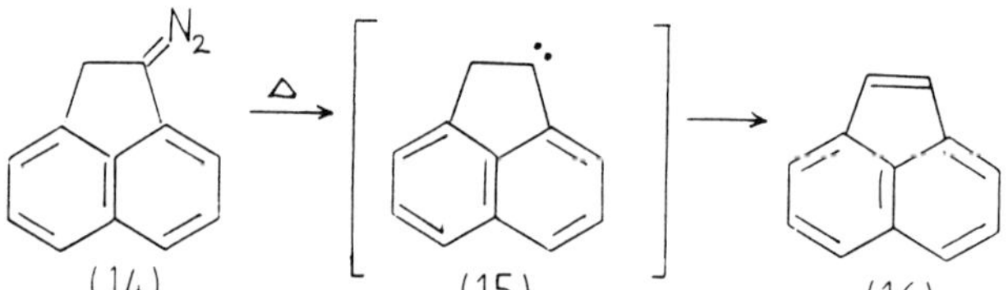

(14) (15) (16)

Reductive electrochemical acetylation of acenaphthylene yields mainly the Z-and E- enol acetates of 1-(1,2-dihydro-1-acenaphthylidene)ethanone, whereas carboxylation followed by methylation gives *trans*-1,2-dimethoxy-carbonyl-1,2-dihydroacenaphthylene (C.Degrand, R.Mora and H.Lund, Acta chem.Scand., 1983, 37B, 429).

The photodimersation of acenaphthylene to *cis*- and *trans*-heptacyclene (2nd Edn, Vol. III, 148), when carried out in micelles, gives an isomer distribution which is affected by substrate concentration, quencher, ionic detergents and heavy atom effects. (H.Mayer and J.Sauer, Tetrahedron Letters, 1983, 24, 4091)

(c) Acenaphthyne

Acenaphthyne **(19)** often postulated as a transient reaction intermediate (see, for example, J.Nakayana *et al.*, Chem. Comm., 1980, 791) has been obtained using a matrix isolation technique (O.L.Chapman *et al.*, J.Amer. chem.Soc., 1981, 103, 7033). Photolysis of the bis(diazo)ketone **(17)** in argon at 15K ($>$416nm.) gives the cyclopropenone **(18)** which on irradiation at 302nm. produces acenaphthyne **(19)** and carbon monoxide.

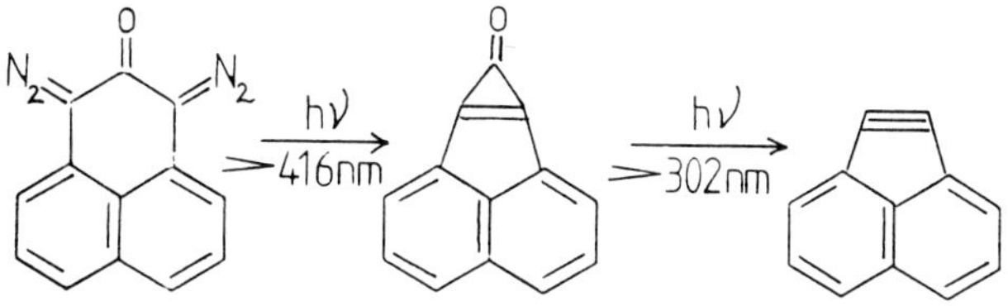

N_2 O N_2 — hν, $>$416nm → — hν, $>$302nm →

(17) (18) (19)

Evidence for the structure of **(19)** rests more securely on its chemical reactions than on its spectroscopic properties, the C≡C triple bond absorption being tentatively assigned to the 1930 cm^{-1} peak in its infrared spectrum. Acenaphthyne reacts with oxygen in the matrix to give acenaphthenequinone, and with a trace of water to give acenaphthenone. Warming **(19)** to room temperature yields decacyclene **(20)**.

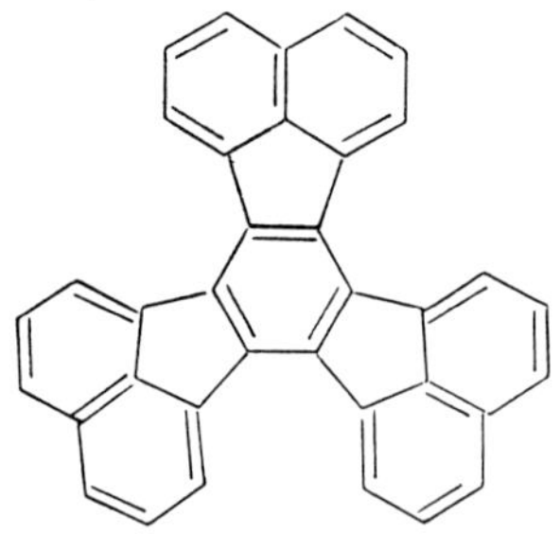

(20)

(d) Benzologues of acenaphthene

(i) Aceanthrene and aceanthrylene

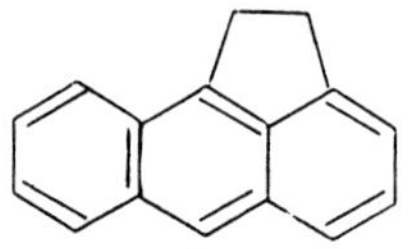

(21) 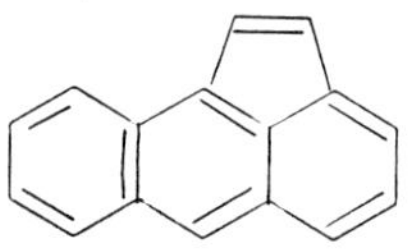 (22)

Although aceanthrene **(21)** is a well-known compound, its parent aceanthrylene **(22)** has been reported only recently. Several related syntheses have been devised for both hydrocarbons. (B.F.Plummer, Z.Y.Al-Saigh and M.Arfan, J.org.Chem., 1984, 49, 2069; H-D.Becker, L.Hansen and K.Andersson, J.org.Chem., 1985, 50, 277; R.Sangaiah and A Gold, Org.prep.Proced.Int., 1985, 17, 53). All the syntheses proceed from anthracene using various acylating agents, and have as key intermediates the isomeric ketones **(23)** and **(24)**.

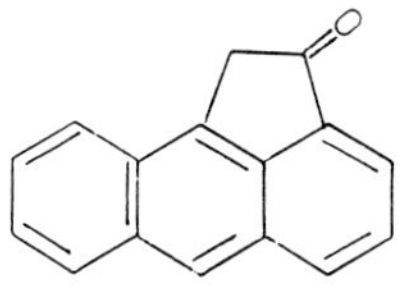

(23) 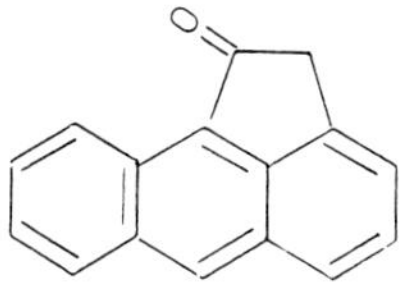 (24)

Conventional reduction and dehydration procedures give **(21)** and **(22)**. There is disagreement as to the m.p. of aceanthrylene: values of 95-96° and 103-104° (scarlet flakes) are given, the latter possibly being the more reliable.

A further discrepancy lies in the reported anomalous fluorescence of **(22)** (Plummer *et al.*) which was not observed by Becker's group.

(ii) Acephenanthrene and acephenanthrylene

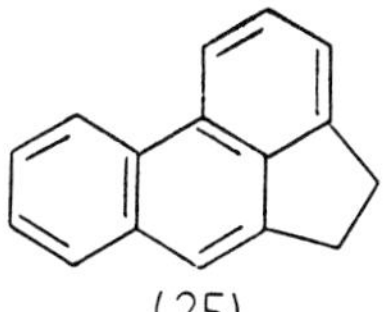

(25) 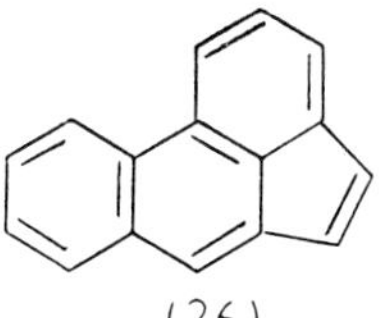 (26)

Acephenanthrylene, cyclopenta[jk]phenanthrene **(26)** m.p. 141-142°, yellow/red crystals (depending upon cooling rate), has been synthesised by a modification of a route originally devised for the known acephenanthrene **(25)**. Thus 5-acenaphthyl-4-butanoic acid is cyclised to the ketone **(27)**, reduced, dehydrated and finally aromatised to **(26)** (L.T.Scott, G.Reinhardt and N.H.Roelofs, J.org.Chem., 1985, 50, 5886).

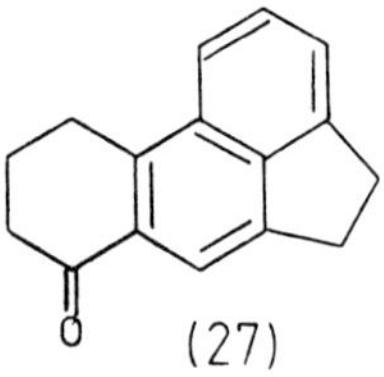

(27)

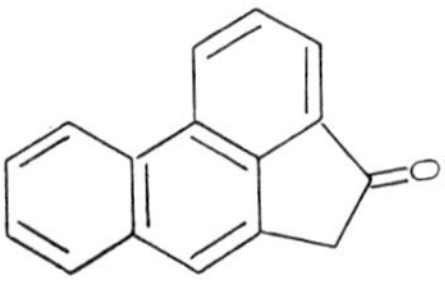

(28)

Alternatively, 9-phenanthryl-2-ethanoic acid, itself formed by elaboration of 9-methylphenanthrene using conventional methods, is cyclised by aluminium chloride to the ketone **(28)**. Reduction of **(28)** followed by dehydration produces **(26)** in 30% overall yield (S.Amin. *et al.*, J.org.Chem., 1985, 50, 4642).

Another synthesis of **(26)** has been described, but with few details (S.Krishnan and R.A.Hites, Anal.Chem., 1981, 53, 342), and a further method is given by G.Neumann and K. Mullen (Chimia, 1985, 39, 275)

(iii) Cholanthrene and its derivatives

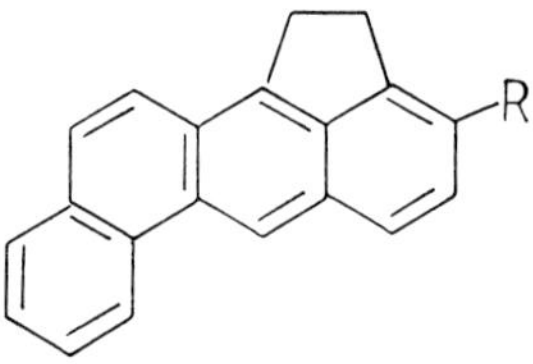

(29) R=H

(30) R=Me

The powerful carcinogenicity of cholanthrene **(29)** and its 3-methyl derivative **(30)** has motivated several workers to devise syntheses which are more convenient than the traditional Elbs procedure. (2nd Edn., Vol III, 152; M.S.Newman and V.K.Khanna, J.org.Chem., 1980, 45, 4507).

In a synthesis claimed (S.A.Jacobs and R.G.Harvey, Tetrahedron Letters 1981, 22, 1083) to be operationally simpler than other methods, and applicable to a wide range of other aromatic hydrocarbons (R.G.Harvey, C.Cortez and S.A.Jacobs, J.org.Chem., 1982, 47, 2120) the indanone **(31)** is condensed with the protected lithium reagent **(32)** to give the lactone **(33)**. Subsequent hydrolysis to the corresponding acid, and reductive ring-closure in the presence of acetic anhydride, furnishes 6-acetoxy-3-methylcholanthrene, which is converted into **(30)** using hydrogen iodide in propanoic acid.

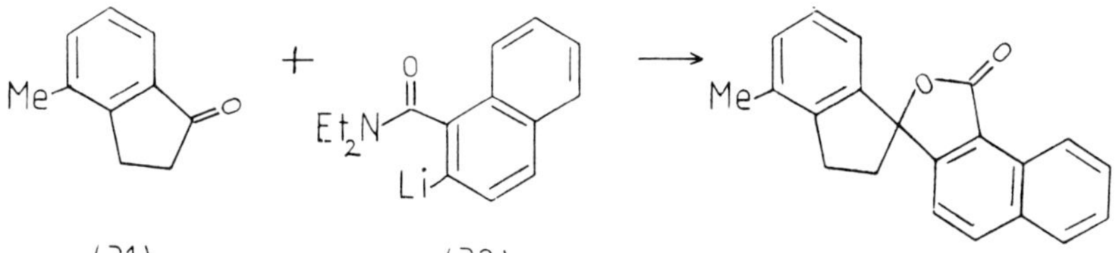

(31) (32) (33)

This route has been applied also to the synthesis of several substituted cholanthrenes in order to test the hypothesis that a methyl group in the 6-position will cause non-planarity and hence enhance carcinogenicity (M.S.Newman and P.K.Sujeeth, J.org.Chem., 1984, 49, 2841). The results obtained suggest that this is indeed the case.

In a neat modification of the synthesis to obtain cholanthrene itself, 2,2-dideuterioindan-1-one is employed to inhibit enolisation of the C=O group. The last step (HI reduction) results in concomitant deuterium removal by exchange (R.G.Harvey and C.Cortez, J.org.Chem., 1987, 52, 283).

Another improved synthesis of 3-methylcholanthrene involves the provision of a better route to the indan derivative **(34)**, used to obtain the key intermediate **(35)** for the conventional Elbs method (P.W.Tang and C.A.Maggicelli, J.org.Chem., 1981, 46, 3429).

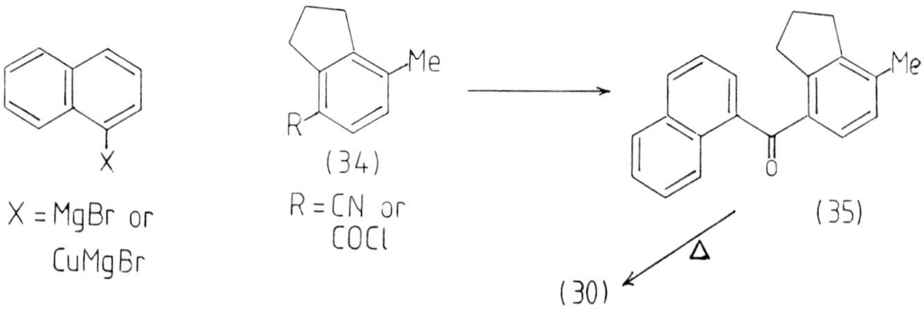

X = MgBr or CuMgBr

(34) R = CN or COCl

(35)

Δ

(30)

The synthesis of **(34)** utilises a Diels-Alder cycloaddition of sorbonitrile to an enamine, followed by aromatisation.

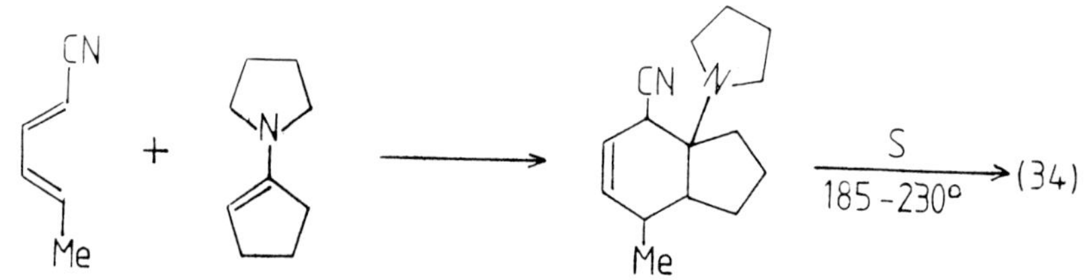

CN

Me

+

N

CN

N

Me

S

185 - 230°

(34)

(iv) Benzaceanthrylenes

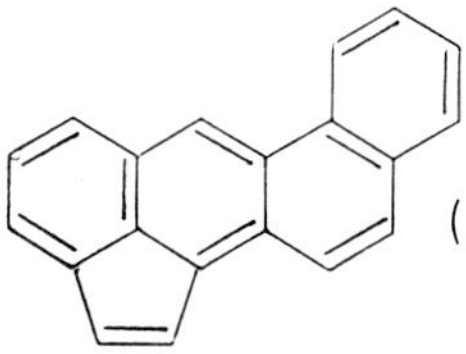

(36) 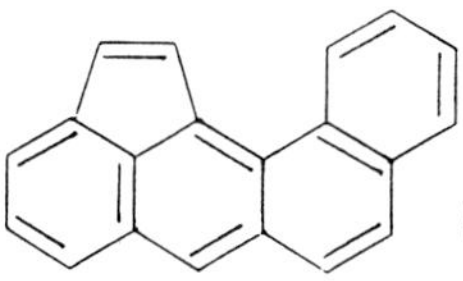 (37)

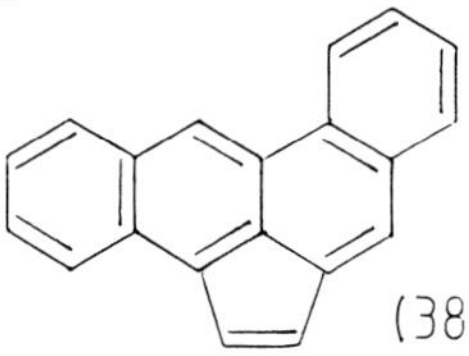

(38) 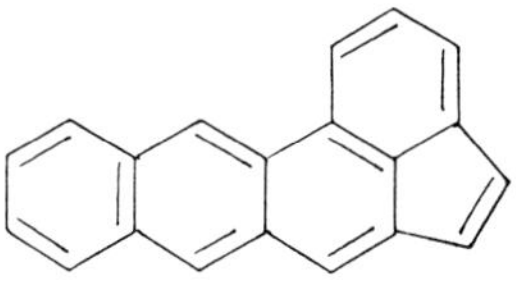 (39)

(36) Benz[j]aceanthrylene, m.p. 170-171°
(37) Benz[l]aceanthrylene, m.p. 157-158°
(38) Benz[e]aceanthrylene, m.p. 138°
(39) Benz[k]aceanthrylene, m.p. 233-234°

All four isomers, although previously synthesised by long routes in low yields, have now been prepared more directly (R.Sangaiah, A.Gold and G.E.Toney, J.org.Chem., 1983, 48, 1632) by cyclisation of the appropriate benzanthryl ethanoic acid, e.g. **(40)**, using hydrofluoric acid. The resulting ketone, e.g. **(41)**, is reduced to an alcohol which is dehydrated in the usual way.

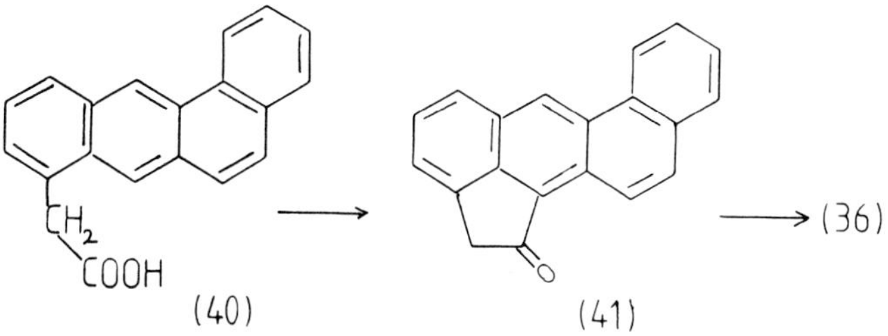

CH_2 COOH (40) → (41) → (36)

(v) *Cyclopenta[cd]pyrene*

Cyclopenta[cd]pyrene **(44)**, $C_{18}H_{10}$, m.p. 176°C, a hitherto neglected hydrocarbon, has been the subject of much recent investigation following the discovery that it is a fairly widespread and potent mutagen, despite the absence of a "bay region".*

The majority of the syntheses of **(44)** depend upon the cyclisation of 4-pyrenylethanoic acid **(42)** to the ketone **(43)**. The corresponding 1-acid is apparently resistant to cyclisation (A.Gold, J.Schultz and E.Eisenstadt, Tetrahedron Letters 1978, 46, 4491; Y.Ittah and D.M.Jerina, *ibid.*, 1978, 46, 4495; P.H.Ruehle, D.L.Fischer and J.C.Wiley, Chem.Comm. 1979, 302).

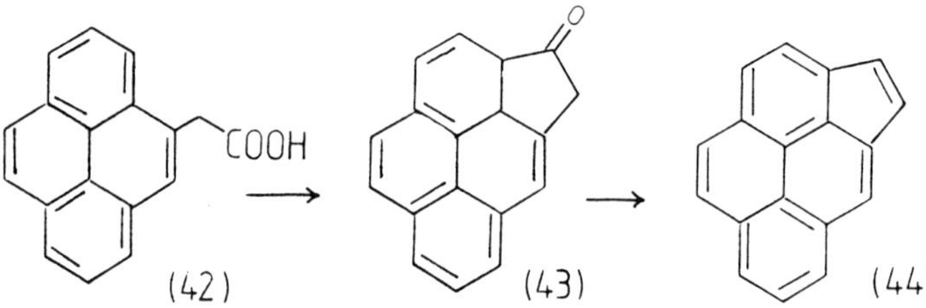

COOH → O →

(42) (43) (44)

A considerable improvement in these syntheses may be effected by taking advantage of a new direct route to the key acid **(42)** (K.Tintel, J.Lugtenburgh and J.Cornelisse, Chem.Comm., 1982, 185). Pyrene is converted into its dianion which on reaction with sodium iodoethanoate gives **(42)** directly. An alternative synthesis of **(42)** makes use of the Pummerer rearrangement (J.C.Wiley *et al.*, J.org. Chem., 1987, 52, 1355)

---

* A "bay region" is a concave exterior region of a polycyclic aromatic hydrocarbon bordered by three benzene rings, at least one of which is a terminal ring.

A more direct route to cyclopenta[cd]pyrene, but giving a low yield, starts with the acetylation of pyrene. The resulting ketone is reduced to the alcohol which on pyrolysis at 850° gives **(44)** in 22% yield (J. Jacob and G.Grimmer, Chem.Abs., 1978, 89, 6143).

Cyclopenta[cd]pyrene is too acid-sensitive to be nitrated by conventional nitrating reagents, but is successfully nitrated using oxidatively-initiated nucleophilic substitution. Treatment of **(44)** with silver nitrate, sodium nitrite and iodine in acetonitrile gives 4-nitro-cyclopenta[cd]pyrene, which is itself sensitive to light, heat and moisture (A.M.van den Braken-van-Leersum, J.Cornelisse and J.Lugtenburg, Tetrahedron Letters, 1985, 26, 4823).

The suspected ultimate carcinogenic metabolite of **(44)**, its 3,4-oxide, has been synthesised in order to determine its activity (A.Gold, J.Brewster and E.Eisenstadt, Chem. Comm., 1979, 903; D.McCaustland, P.H.Ruehle and J.C. Wiley, *ibid.*, 1980, 93).

3 The fluorene group

(a) Fluorene

(i) Methods of formation *

It is perhaps a testimony to the thoroughness with which fluorene has been investigated that few novel syntheses have been reported since the publication of C.C.C., 2nd Edn.

One synthetic route which does merit attention for its interesting mechanism and generality of application is the rearrangement of α-alkoxycarbonyldiarylmethyl cations **(45)** to fluorene-9-carboxylic esters **(46)** (E.Lee-Ruff *et al.*, Chem.Comm., 1983, 727 ; A.C.Hopkinson, E.Lee-Ruff and M.Maleki, Synthesis, 1986, 367).

* A Review of the construction of the fluorene skeleton is given by K. Suzuki and M. Minabe in Yuki Gosaku Kyokaishi, 1981, 39, 122.

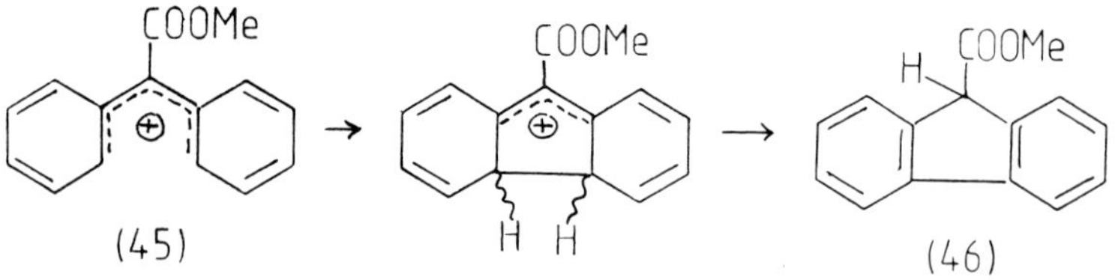

Conventional hydrolysis and decarboxylation procedures provide fluorene and its derivatives. The requisite cations (**45**) are prepared by the reaction of an aryl Grignard reagent with $PhCOCOOCH_3$. The resulting hydroxy-ester gives (**45**) on treatment with sulphuric acid at $< 10^{o}C$.

(ii) ***Electrophilic substitution***

In a systematic study of the acetylation of substituted fluorenes (D.R. Buckle, N.J. Morgan and R.G. Alexander, J.chem.Soc., Perkin I, 1979, 3004) it has been demonstrated that the nature and position of some substituents have a profound influence on the course of further substitution (Table 1).

TABLE 1

Substrate	Diacetyl product	Yield
Fluorene	2,7-	72
1-Methylfluorene	2,7-	66
1-Methoxyfluorene	4(?),7- ; 2,4-	*
3-Bromofluorene	2,7- ; 1,7-	94, trace
4-Fluorofluorene	1,7-	37

* some dealkylation occurs

Thus the C-1 and C-4 positions can be selectively acylated even in situations where C-2 is vacant. It appears that control of substitution involves a balance between the high electron densities at C-2 and C-7, and the directive effect of the substituent.

A useful technique using blocking groups has been developed to synthesise 4-substituted fluorenes, hitherto obtainable only by a tedious route from fluorene-4- carboxylic acid (S.Kajigaeshi *et al.*, Synthesis, 1984, 4, 335; Bull.chem.Soc.Japan, 1986, 59, 97).

Fluorene is first converted into 2,7-di*tert*butyl fluorene using 2,6-di*tert*butyl-*p*-cresol and aluminium chloride in nitromethane. The bulky *t*-butyl groups prevent further substitution in the contiguous positions. The following transformations are then effected regioselectively.

$Br_2/Fe \longrightarrow$ 4-bromo

$I_2/HIO_4/H_2SO_4 \longrightarrow$ 4-iodo

$ZnCl_2/CH_3OCH_2Cl \longrightarrow$ 4-chloromethyl

HNO_3(fuming)/Ac_2O/HOAc $\longrightarrow$ 4-(and 7-) nitro

The above substituents may then undergo further transformations to provide the functions $-CH_3$, $-CH_2COOH$, $-CH_2OH$, $-NH_2$, -NHAc. Finally, the products are de-*t*-butylated by heating them with $AlCl_3$ in benzene, but this step proves to be limited to substituents with high electron densities. The reaction is most successful with -Br, $-CH_3$ and -NHAc substituents.

An alternative route to 4-substituted fluorenes or fluorenones (D.Hellwinkel and G.Haas, Ann., 1978, 1913) employs nitro groups in the 2- and 7- positions as blocking groups. These are subsequently removed by reduction, diazotisation and treatment with hypophosphorous acid. 4,5-Dibromofluorenone and 4,5-diiodofluorenone are readily available by this route.

(iii) 9-Substituted fluorenes

A vast literature exists in this area on topics largely irrelevant to the present review, the role of the fluorene nucleus being simply that of a stable locus for a reactive methylene group. Fundamental aspects of the chemistry of such compounds are well covered in the 2nd Edition, Vol.IIIH, pp.160-161, 168-175.

The stable ylid, triphenylphosphonium fluorenylide **(47)** has received considerable attention. For example, with diphenylketene and diphenylethanal it gives respectively **(48)** and **(49)** (M.I.Shevchuk, V.N.Kushnir and A.V.Dombrovskii, Chem.Abs., 1978, 88, 89425).

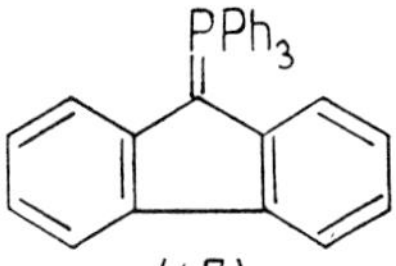

(47)

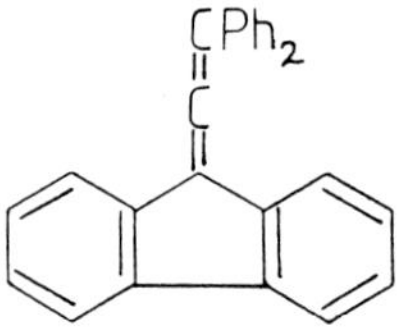

(48)

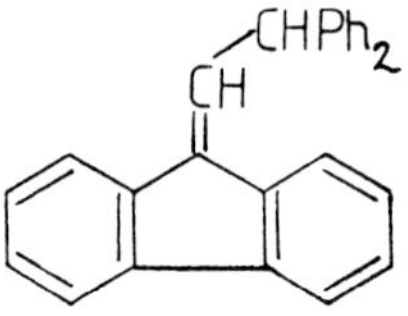

(49)

Reaction with *N*-sulphinyl-*p*-toluene-sulphonamide occurs at the S=O bond, but with *N*-sulphinyl-*p*-nitroaniline the N=S bond is attacked (T.Saito and S.Motoki, J.org.Chem., 1977, 42, 3922).

The related phosphonate carbanions e.g. **(50)** have been prepared and found to react with carbonyl compounds in the usual way to give alkenes in 50-90% yields (R.S.Tewari, N.Kumari and P.S. Kendurkar, J.Indian chem.Soc., 1977, 54, 443).

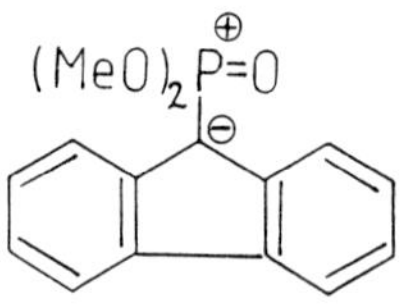

(50)

The dinitro-9-methylenefluorene **(51)** provides some interesting chemistry (S.Hoz and D.Speizman, Tetrahedron Letters, 1979, 50, 4855). With most common nucleophiles ($Nu^- = CN^-$, OMe^-, $PhCh_2S^-$) stable carbanions having the structure **(52)** are obtained, but sodium azide yields the unexpected product *N*-cyano-9-iminofluorene **(53)**.

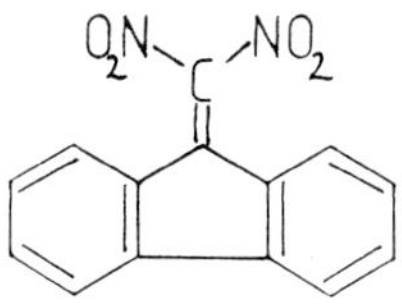

(51) 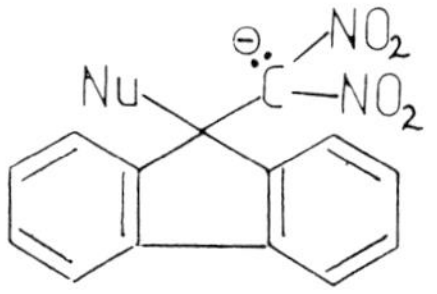 (52) 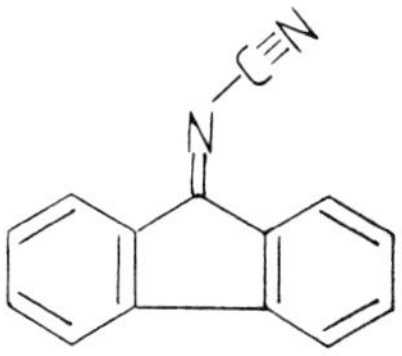 (53)

9,9-Dichlorofluorene, when treated with sodium triphenylmethyl, gives rise to the 9-tritylfluorenyl radical which has been characterised by its esr spectrum. (W.B.Smith and M.C.Harris, J.org.Chem., 1983, 48, 4957).

Side-chains on the 9-position of fluorene are conventionally obtained *via* reactions of the fluorene anion. An unusual alternative is the coupling of fluorenone with enones in the presence of $TiCl_4$-Mg to give γ-ketol derivatives, presumably by conjugate addition of the fluorenone dianion (J-M.Pons and M.Santelli, Tetrahedron Letters, 1982, 23, 4937).

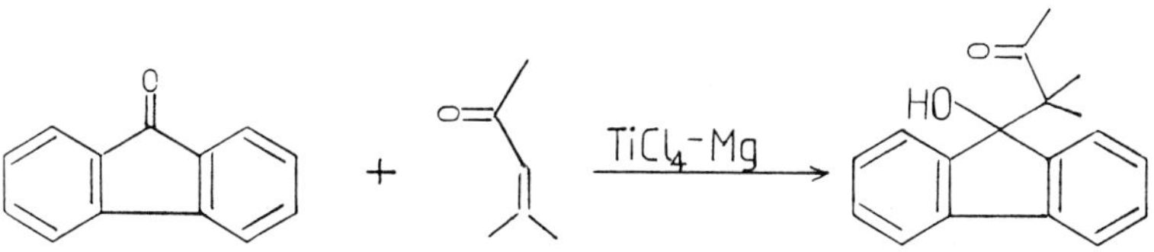

$TiCl_4$-Mg

Cathodic reduction of 9,9-dicyanofluorene and related compounds produces dianions (B^{2-}) which efficiently convert phosphonium salts into ylids (R.R.Mehta, V.L. Pardini and J.H.P.Utley, J.chem.Soc., Perkin I., 1982, 2921).

$$R\text{-}CH_2\text{-}\overset{\oplus}{P}Ph_3 \longrightarrow R\text{-}CH{=}PPh_3 \xrightarrow{R'CHO} R\text{-}CH{=}CH\text{-}R'$$

$$B \rightleftharpoons B^{2\ominus} \quad BH^{\ominus}$$

Advantages of cathodically-generated bases include: convenience of stable and easily-stored probases, control over generation rate, reproducibility of reaction conditions and possibility of stereochemical control by variation of cation. One notable result is the formation of Vitamin A acetate containng 76% of the 11-*cis* isomer.

Another application of fluorene derivatives in synthesis involves the surprising efficacy of 9-fluorenone-1-carboxylic acid in transamination reactions (C.A.Panetta and A.S.Dixit, J.org.Chem., 1980, 45, 4503).

$$\text{9-fluorenone-1-COOH} + R\text{-}CH(Y)\text{-}NH_2 \xrightarrow[\text{Toluene},\ \Delta]{\text{HOAc}} \text{9-}\overset{\oplus}{N}H_3\text{-fluorene-1-}COO^{\ominus} + R\text{-}C(Y){=}O$$

(R = alkyl, H
Y = Ph, H)

Disappointingly, aminoacids fail to give the corresponding ketoacid, despite being clearly transaminated.

A feature of certain 9-substituted fluorenes which has attracted some attention is that of steric hindrance. 9-Arylfluorenes are unique systems which exhibit high barriers to rotation about the C-9—C-Ar bond. (M.Ōki, Angew.chem.Intern.Edn., 1976, 15, 87). Thus compounds of the type **(54)** give stable rotamers, sp and ap, at room temperature. The ap conformer is favoured for hydroxylic compounds, whereas the sp is dominant in methyl ethers. (M.Nakamura and M.Ōki, Bull.chem.Soc.Japan 1980 53, 3248; T.Mori and M.Oki, *ibid.*, 1981, 54, 1199).

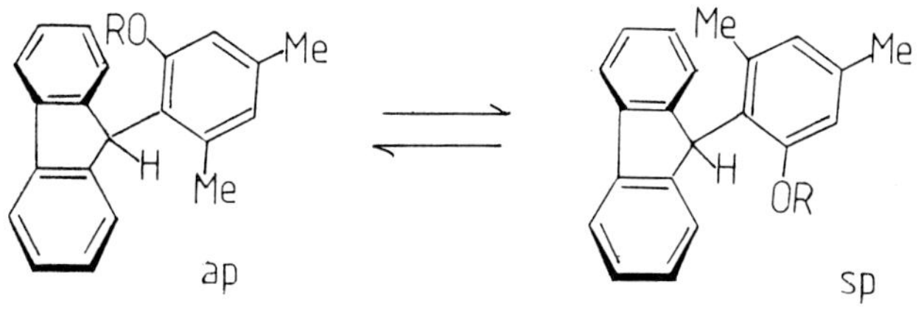

ap (54a) ⇌ sp (54b)

(iv) Fluorofluorenes

Octafluorofluorene, $C_{13}H_2F_8$, m.p.113.5-115°, presents an intriguing comparison with its hydrocarbon analogue (R. Filler, A.E.Fiebig and M.Y.Pelister, J.org.Chem. 1980, 45, 1290), particularly with respect to reactions at the 9-position. The polyfluoroaryl groups strongly stabilise 9-carbanions, causing enhancement of acidity and acceleration of carbanionic rearrangements. In fact, the dichotomy betweeen fluorene and octafluorofluorene is one of the most revealing manifestations of the differences betweeen H and F analogues. Octafluorofluoren-9-one, for example, cannot be reduced to octafluorofluorene; the anion of the latter, generated using sodium hydride, fails to undergo carboxylation or methylation; and the 9-ester resists reduction to the alcohol, the dominant reaction being proton abstraction to give the very stable carbanion. On the other hand, hydrolysis and decarboxylation of the ester is extremely facile, again owing to the exceptional stability of the 9-carbanion.

(v) *Other fluorenes*

The 4aH-fluorene system is of interest because of the wide variety of cycloaddition reactions and sigmatropic rearrangements which it undergoes (D.Neuhaus and C.W. Rees, Chem.Comm., 1983, 318). 4a-Methyl-4aH-fluorene (55), for example, is isolable but very reactive. Its reactions include autoxidation, dimerisation, photochemical rearrangement, [4 + 2] cycloadditions and acid-catalysed aromatisation to a mixture of 1- and 4-methylfluorenes. On flash vacuum pyrolysis at 650°C it gives mainly 9-methylfluorene.

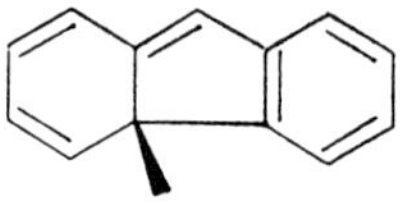

(55)

(b) *Bifluorenes and related compounds*

Bifluorenylidene, $(C_6H_4)_2$ C=C$(C_6H_4)_2$, (2nd Edn., Vol.IIIH, p.175) has been prepared in many ways and is often an unwanted product arising from the dimerisation of fluorene derivatives. A novel coupling procedure to produce bifluorenylidene uses low-valent titanium salts (P.Lemmen and D.Lenoir, Ber. 1984, 117, 2300). The influence of substituents at C-1 and C-8 on the geometry of such compounds is the subject of much interest (Lemmen and Lenoir, *loc.cit.*; M.Ballaster *et al.*, J.org.Chem., 1985, 50, 2287).

Another property of the fluorenylidene group is its considerable charge-spreading ability, leading to the remarkable acidities of certain hydrocarbons containing this moiety, e.g. *bis*biphenylenepropene $(C_6H_4)_2$ C:CH.CH$(C_6H_4)_2$, pK_A=10 (2nd Edn. III H, 176). This hydrocarbon is conveniently prepared by treating fluorene with acetonitrile, potassium hydroxide powder and oxygen (H.Bauer, C.Moerler and D.Rewicki, Ber., 1979, 112, 1473).

Fluorenophanes. The [2.2] phanes derived from fluorene, fluorenone and the 9-fluorenyl anion have been synthesised by conventional procedures (2nd Edn. IIIH, p.177; IIIF, p.347) in order to serve as models for excimers and exiplexes (M.W.Haenel, H.Irngartinger and C.Krieger, Ber., 1985, 118, 144).

(c) Benzofluorenes

The synthesis of fluorenes outlined on p.39 (A.C.Hopkinson, E.Lee-Ruff and M.Maleki, Synthesis, 1986, 367) may, with appropriate modification, be used to prepare benzo[a] and benzo[c]fluorenes, as well as several dibenzofluorenes.

Other, less generally useful, syntheses have been described (M.P.Reddy and G.S.K.Rao, J.org.Chem., 1981, 46, 5371; D.N.Chatterjee and J.Guha, Curr.Sci., 1986, 55, 503).

Benzo[def]fluorene **(56)**, 4H-cyclopenta[def]phenanthrene, has recently been subjected to thorough and systematic study, notably by Yoshida's group*. The main justification for such interest in **(56)** lies in its unique incorporation of the special features of both fluorene and phenanthrene. It also represents a simple model for polycyclic hydrocarbons with "bridged" bay regions.

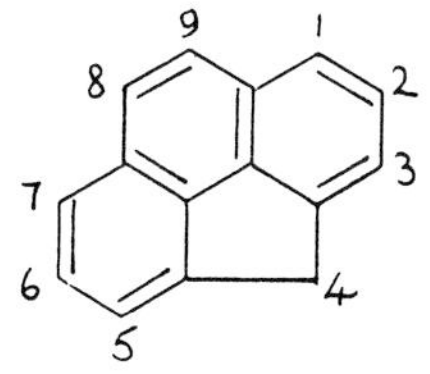

(56)

Synthesis of **(56)** has previously been achieved from the phenanthrene and acenaphthene nuclei (2nd Edn., IIIH,

* Review of the synthesis and reactions of cyclopenta-[def]phenanthrene, M.Minabe and K.Suzuki, Yuki Gosei Kagaku, Kyokaishi, 1986, 44, 421.

pp.180-181), and the third possibility, namely from the fluorene skeleton, has now been realised (M.Yoshida, M.Minabe and K.Suzuki, Bull.chem.Soc.Japan, 1983, 56, 2179). Standard ring-building procedures starting from 4-fluorenylethanoic acid give 8-hydroxycyclopenta[def]phenanthrene (which incidentally provides a route to otherwise inaccessible 8-derivatives), and this is readily converted into **(56)** with hydrogen iodide and phosphorus.

Friedel-Crafts acetylation of **(56)**, previously reported by W.E.Bachmann and J.C.Sheehan (J.Amer.chem.Soc., 1941, 63, 2598), has been repeated (M.Yoshida *et al*., J.org.Chem., 1979, 44, 1783). The products are 1-acetyl (with some 2-, 3- and 8-) when $CHCl_3$ or CH_2Cl_2 is used as solvent, and a mixture of the 1- and the 3- acetyl when the reaction is carried out in nitro-benzene or nitromethane. The "3-acetyl" compound reported by the former workers was clearly a mixture of the 1-, the 2- and the 3- isomers.

Nitration of **(56)** (*idem*., *ibid*., 1979, 44, 1915) yields a mixture of the 1-, the 2-, the 3- and the 8- nitro-compounds. In contrast 8,9-dihydrocyclopenta[def]phenanthrene gives only the 2-nitro-derivative and cyclopenta[def]phenanthren-4-one gives mainly the 8-isomer.

Bromination of **(56)**, the 8,9-dihydroderivative and the 4-one proceeds along similar lines, (*idem*., *ibid*., 1979, 44, 3029), but the relative amount of 1-isomer is much higher than for nitration, consistent with the generally observed higher regioselectivity in bromination. Likewise, sulphonation with sulphuric or chlorosulphonic acid (*idem*., Bull.chem.Soc.Japan, 1981, 54, 1159) affords mainly the 1-sulphonic acid, the 8,9-dihydroderivative gives the 2-acid, and the 4-one yields predominantly the 8- acid. The acids are conveniently separated as their esters by hplc.

The melting points of several monosubstituted cyclopenta[def]phenanthrenes are collected in Table 2.

TABLE 2

Monosubstituted cyclopenta[def]phenanthrenes

Substituent	Position	m. p. (°C)
$-COCH_3$	1	156 - 157
	2	128.5 - 129
	3	113.5 - 115
	8	94.5 - 95.0
$-NO_2$	1	175.5 - 176.5
	2	172 - 173
	3	153 - 154
	8	188 - 189
$-NH_2$	1	128 - 129
	2	126 - 127
	3	91.5 - 92.5
	8	136 - 138
NHac	1	216.5 - 217
	2	227 - 228.5
	3	227.5 - 229 dec.
	8	230 - 231
-Br	1	85.5 - 86.5
	2	91 - 92
	3	81 - 82
	8	98 - 99
	4 (Bridge)	130 - 132

The rate coefficients for each position in **(56)** in TFA at 70° have been determined by protiodetritiation (W.J. Archer and R.Taylor, J.chem.Soc., Perkin II, 1981, 1153). These are: 1-, 27050; 2-, 5680; 3-, 1400; 8-, 6950, with corresponding σ^+ values of -0.5065, -0.43, -0.475 and -0.44. The results confirm that **(56)** possesses a combination of the properties of fluorene and phenanthrene, and the relative positional reactivities correlate well with Huckel localisation energies. The σ^+ values correctly predict the isomer distribution in nitration, but less satisfactorily those for acetylation and bromination. It is interesting to observe that the reactivity of the 8- position is six times less than that of the corresponding position (9-) of (non-planar) 4,5-dimethylphenanthrene. The reactivity of the former is quite satisfactorily explained by the presence of the CH_2 bridge (i.e. similar to that of 4-methylphenanthrene). It follows that 4,5-dimethylphenanthrene must be unusually reactive - due no doubt to its non-planarity which destabilises the ground state (H.V.Ansell and R.Taylor, J.org.Chem., 1979, 44, 4946).

Disubstitution in cyclopenta[def]phenanthrene has received scant attention to date. Some pertinent results are presented in Table 3, (M.Yoshida *et al*., Bull.chem.Soc. Japan, 1983, 56, 1259).

TABLE 3

Bromination of substituted cyclopenta[def]phenanthrenes

Position occupied	Substituent	Position of bromination
1	NO_2, Br, CH_3CO	5 or 7
2	"	"
3	"	
1	NH_2, NHAc	2, then 8
2	"	1, then 3
3	"	8, then 2
8	"	9, then 3

2,6,8-Trinitro-4H-cyclopenta[def]phenanthren-4-one (TNC) has been synthesised, and found to be a more effective complexing agent than trinitrofluorenone (TNF) (M.Minabe, M.Yoshida and O.Kimura, Bull.chem.Soc.Japan, 1985, 58, 385).

Catalytic hydrogenation of **(56)** in the presence of Raney nickel gives a mixture of the 8,9-dihydro- and the 1,2,3,3a-tetrahydro-derivatives (M.Minabe *et al*., Bull.chem.Soc.Japan, 1984, 57, 725), each of which is reduced further according to the following scheme.

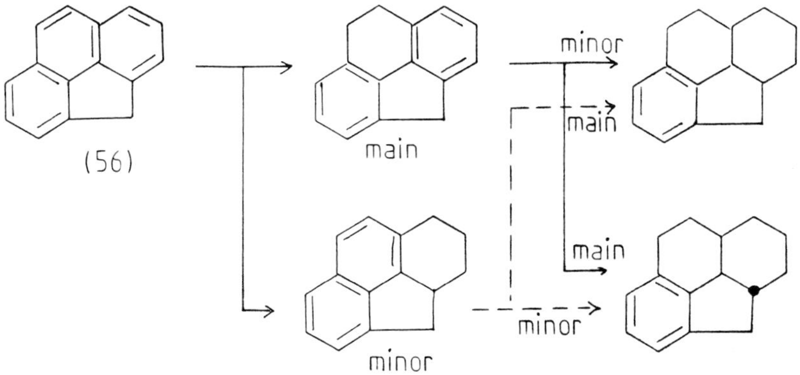

Oxidation of **(56)** is discussed in the 2nd Edn. Vol.IIIH, p.181. The 8,9-dihydro derivative may be oxidised more selectively. Trivalent chromium yields the 8,9-quinone, whereas Triton B-oxygen gives the 4-one (M.Yoshida *et al*., Bull.Chem.Soc.Japan, 1980, 53, 1179).

Ozone attacks **(56)** predictably at the phenanthrene-type 8,9-bond to give a stable monomeric ozonide (M.Yoshida *et al*., Tetrahedron, 1979, 35, 2237; I.Willner and M. Rabinovitz, Org.Prep.Proced.Intern., 1980, 12, 351). Fluorene-4,5-dialdehyde, diol, diacid, lactone, acid anhydride and aldehyde-acid are obtained from the ozonide in the conventional ways.

Like 9-hydroxymethylfluorene, which undergoes Wagner-Meerwin rearrangement to phenanthrene (2nd Edn. IIIH, 170), 4-hydroxymethylcyclopenta[def]phenanthrene rearranges to give pyrene (T.Kimura, M.Minabe and K. Suzuki, J.org.Chem., 1978, 43, 1247).

Related Compounds

15,16-Dihydro-1,11-methanocyclopenta[a]phenanthren-17-one **(57)**, a carcinogen with a bridged bay region, has been synthesised, its structure determined by X-ray diffraction, and its biological activity assessed (M.M. Coombs *et al.*, Chem.Comm., 1979, 433).

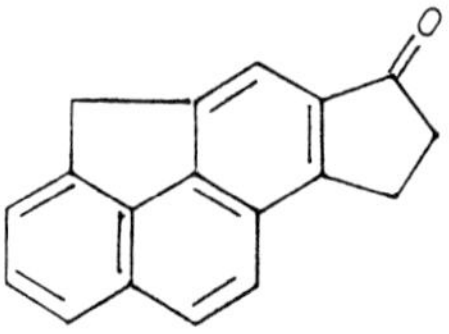

(57) 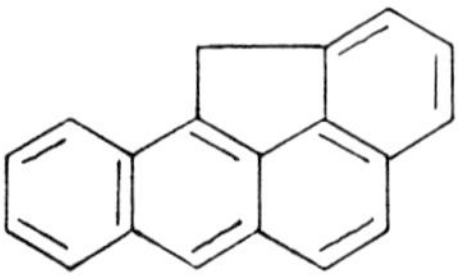 (58) 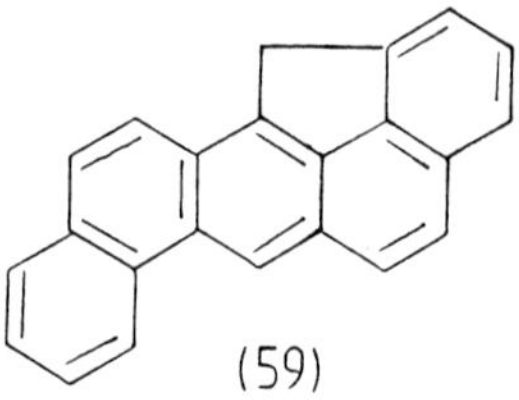 (59)

With the object of preparing examples of methylene-bridged polycyclic hydrocarbons, which are relatively rare, J.K.Ray and R.G.Harvey (J.Org.chem., 1983, 48, 1352) have applied the *o*-lithioaryl amide route (p.35) in an elegant manner to the synthesis of **(58)** and **(59)**.

(d) Higher annelated fluorenes

A helicene containing the fluorene nucleus, phenanthro [3,4-c]fluorene **(60)** has been prepared with a view to studying the effect of a 5-membered ring on the helical conformation by comparison of nmr data with those of hexahelicene. (J.W.Diesveld, J.H.Borkent and W.H. Laarhoven, Tetrahedron, 1982, 38, 1803). Synthesis of **(60)** is accomplished as shown below.

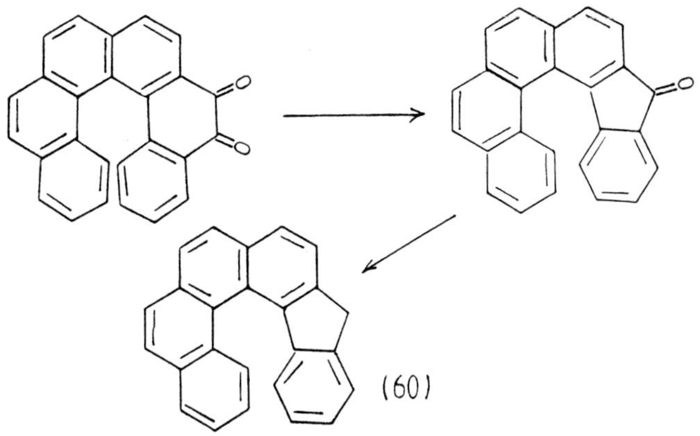

(60)

(e) Natural products of the fluorene group

Compounds containing the fully-aromatic fluorene nucleus (c f. 2nd Edn., IIIH, p.183) have been isolated from the orchid *Dendrobium gibsonii* (S.K.Talapatra *et al.*, Tetrahedron, 1985, 41, 2765).

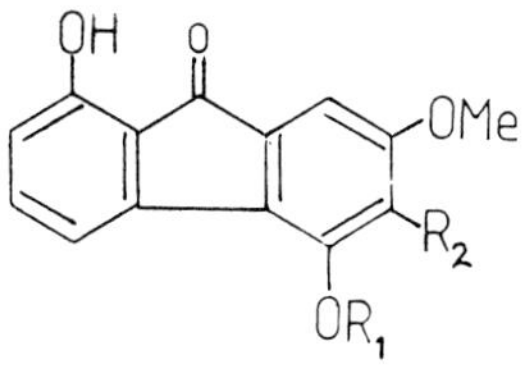

$R_1 = R_2 = H$: Dengibsin

$R_1 = Me$, $R_2 = OH$: Dengibsinin

These are the first known naturally occurring fluorenones.

4 The fluoranthene group

A review by N.Campbell and N.H.Wilson (Comm.Roy.Soc.Edin. (Phys.Sci.), 1979, 15, 193, together with 2nd Edn., IIIH, pp.184-197, represents the current state of the fundamental chemistry of fluoranthene and its derivatives. A selection of its more recent syntheses and reactions are given in the sequel.

(a) Fluoranthene

(i) Methods of formation

The "meneidic" property of aromatic hydrocarbons, that is their tendency to "retain the type" is amply illustrated by the transformations of the diverse compounds **(61-65)** into fluoranthene or one of its derivatives.

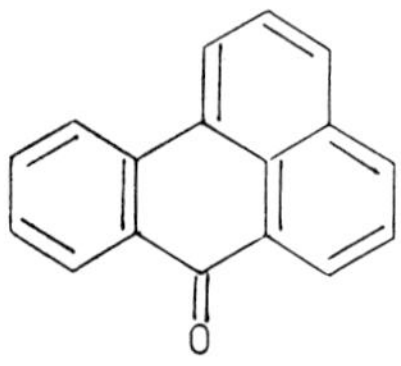

(61)

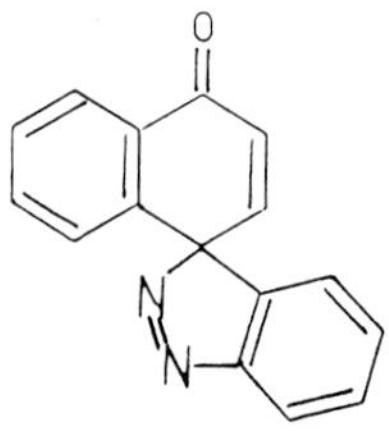

(62)

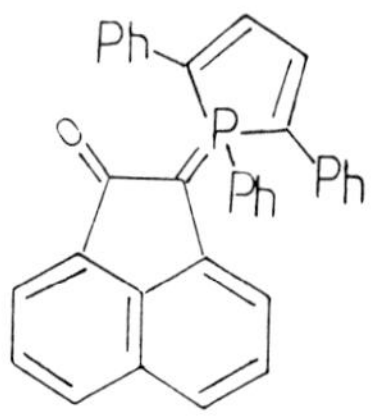

(64)

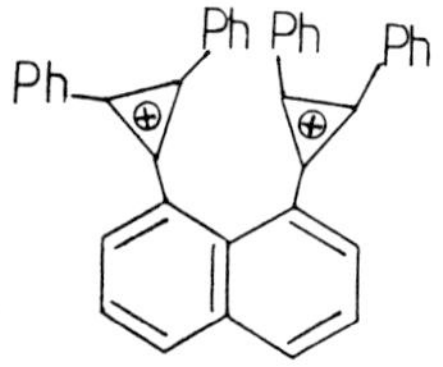

(63)

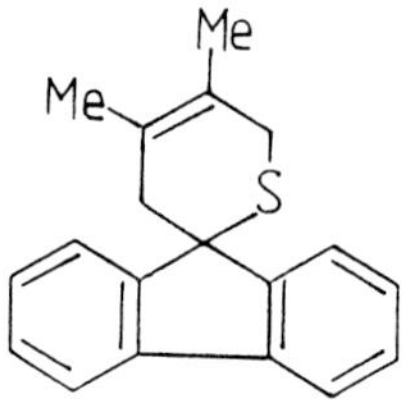

(65)

Benzanthrone **(61)** (E.Maly, Monatsh.Chem., 1981, 112, 1103) and the spiroindazole **(62)** (K.Hirakawa *et al*., J.chem.Soc., Perkin I, 1980, 1944) each give fluoranthene on heating with zinc dust. The unusual *bis*(cyclopropylium) salt **(63)** gives 7,8,9,10-tetraphenylfluoranthene upon treatment with the same reagent (K.Komatsu, M.Arai and K.Okamoto, Tetrahedron Letters, 1982, 23, 91), while thermolysis of the phospholylidene-acenaphthenone **(64)** yields 7,10-diphenylfluoranthene (J.I.G.Cadogan, A.G. Rowley and N.H.Wilson, Ann., 1978, 74). Irradiation of the spirodihydrothiopyran **(65)** produces 2,3-dimethylfluoranthene (K.Praefcke and Ch. Weichsel, Tetrahedron Letters, 1976, 1787).

Fluoranthene is also formed by reductive alkylation of acenaphthylene with 1,4-dichlorobutane using lithium and ammonia (G.Neumann and K.Muellen, Chimia, 1985, 39, 275).

A reinvestigation of previous work by S.H.Tucker (H.G. Heller and G.A.Jenkins, J.chem.Soc., Perkin I, 1984, 2871) has shown that the fluorenylidene derivative **(66)**, on iradiation in toluene, cyclises to **(67)**, which in turn undergoes a thermal suprafacial 1,9-H shift to give **(68)**.

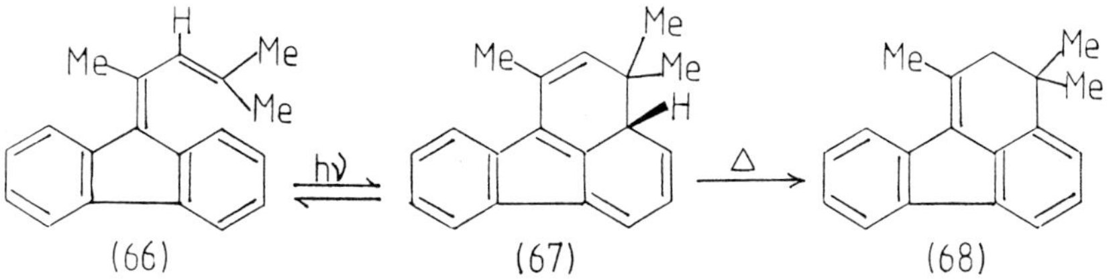

(66) (67) (68)

The use of more conventional routes to fluoranthene derivatives is well illustrated by the synthesis of all the possible fluoranthenols (Table 4) together with *trans*-2,3-dihydroxy-2,3-dihydrofluoranthene **(69)** which is thought to be the major mutagenic metabolite of fluoranthene (J.E.Rice, R.J.La Voie and D.Hoffmann, J.org.Chem., 1983, 48, 2360; W.H.Rastetter *et al.*, *ibid.*, 1982, 47, 4873)

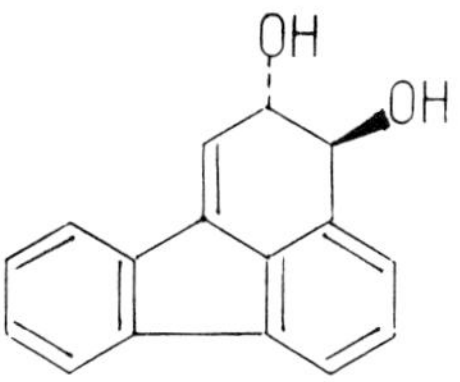

(69)

TABLE 4

Fluoranthenols

Isomer	m.p.(^{o}C)
1-	149 - 150
2-	141 - 141.5
3-	186.5 - 187.5
7-	154 - 157
8-	155 - 157.5 (Lit.162)

A new synthesis of fluoranthene derivatives employs cyclimonium salts such as **(70)** as "masked" αβ-unsaturated ketones (J.Curtze *et al.*, Ber., 1979, 112, 2190).

COR
CH
CH
R
NaOMe
DMF
H
COR
H
H
R
(70)
R
CH_2COR
H
$R\diagdown C \diagup CH_2COR$
R
R

(ii) Chemical reactions

Few significantly new reactions of fluoranthene have been reported since the 2nd Edition, but the following merit attention: A stable radical cation, [fluoranthene]$_2^{\odot\oplus}$ $PF_6^{\ominus}$ is generated by anodic oxidation in dichloromethane at -30°C (C.Krohnke, V.Enkelmann and G.Wegner, Angew.Chem.intern. Edn., 1980, 19, 912). Such radical cations, also formed from other aromatic hydracarbons, constitute a new family of "organic metals" in which the mobility of charge carriers is strongly temperature dependent.

Fluoranthene can be methylated by nucleophilic addition of an alkyllithium, followed by treatment with iodine (D.A.Peake *et al.*, Synth.Comm., 1983, 13, 21).

Reaction of fluoranthrene with cobalt fluoride gives perfluoro-perhydrofluoranthene which is defluorinated over iron[III] oxide at 420° to decafluorofluoranthene (J.Burdon *et al.*, J.chem.Soc., Perkin I, 1980, 1726).

(b) Benzofluoranthenes

Benzo[a]fluoranthene (72) is formed by cyclodehydrogenation of 9-phenylanthracene using platinum on charcoal at 400°C (P.Studt, Ann., 1979, 1443). More usefully, it is synthesised by R.G.Harvey's *o*-lithioarylamide route (p.35) using fluorenone as the substrate (J.K.Ray and R.G. Harvey, J.org.Chem., 1982, 47, 3335). It is noteworthy that the last stage in this synthesis, reductive cyclisation of (71) with hydrogen fluoride, occurs in a single step, and is promoted by the presence of a hydride-ion donor such as triphenylmethane.

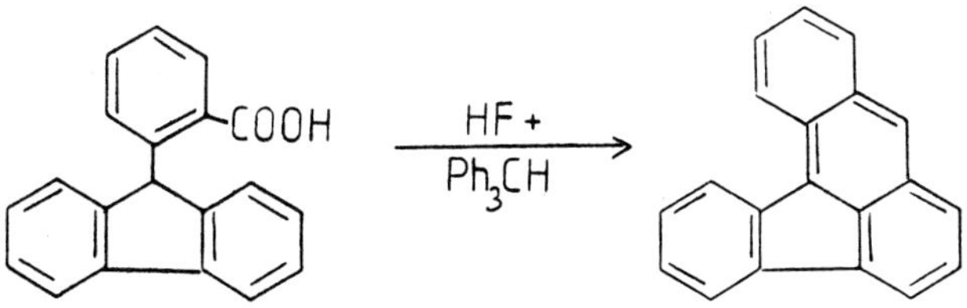

(71) (72)

Benzo[b]fluoranthene, owing to its proven carcinogenicity, has received considerable attention. In an intensive programme, S.Amin and co-workers have synthesised all twelve isomeric hydroxy-derivatives (J.org.Chem., 1986, 51, 1206) (Table 5) and several methyl-derivatives (*ibid.*, 1985, 50, 1949).

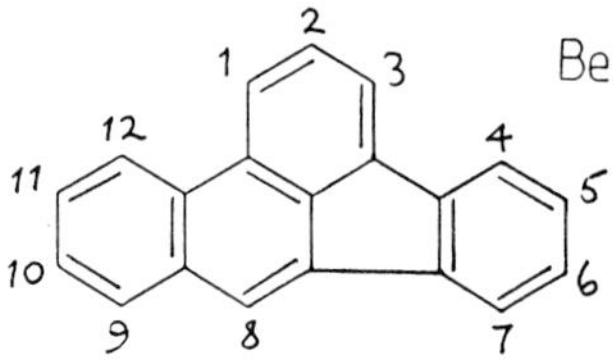

Benzo[b]fluoranthene

TABLE 5

Hydroxybenzo[b]fluoranthenes

Isomer	m.p.	Isomer	m.p.(°C)
1-	235 - 236	7-	*
2-	221 - 222	8-	209 - 210
3-	248 - 249	9-	221 - 222
4-	*	10-	219 - 230
5-	*	11-	215 - 216
6-	232 - 233	12-	214 - 215

* spectroscopic quantities only

The dihydro-diol metabolites **(73)** and **(74)** have been synthesised (*idem.*, *ibid.*, 1984, 49, 1091) and the absolute configuration of each enantiomer of **(73)** established (S.K.Yang, M.Mushtaq and L.Kan, *ibid.*, 1987, 52, 125).

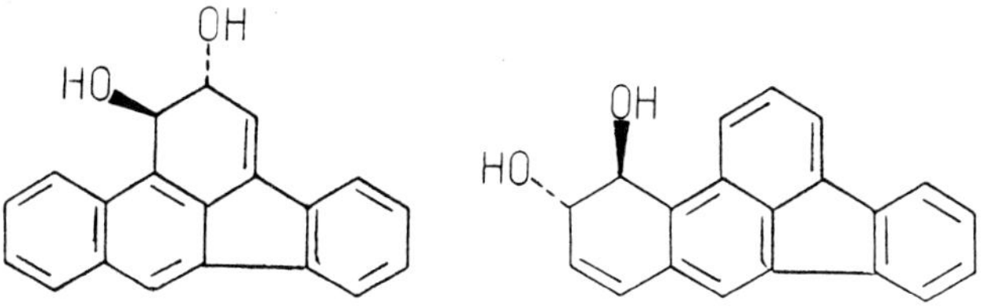

(73) (74)

Benzo[j]fluoranthene, also a carcinogen, has been prepared by the following route (E.J.Eisenbraun *et al.*, J.org.Chem., 1984, 49, 1030).

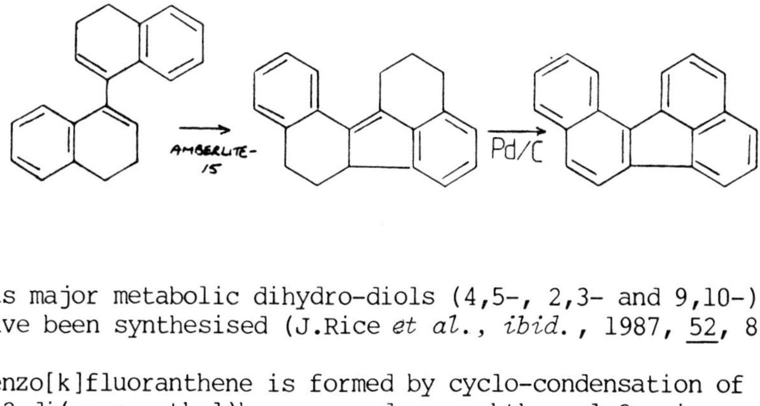

AMBERLITE-15 Pd/C

Its major metabolic dihydro-diols (4,5-, 2,3- and 9,10-) have been synthesised (J.Rice *et al.*, *ibid.*, 1987, 52, 849)

Benzo[k]fluoranthene is formed by cyclo-condensation of 1,2-di(cyanomethyl)benzene and acenaphthene-1,2-quinone, followed by decyanation and acidification (E.H.Vickery and E.J.Eisenbraun, Org.Prep.Proced.Intern., 1979, 11, 259).

Benzo[ghi]fluoranthene can be obtained directly from benzo[c]phenanthrene (P.Studt and T.Win, Ann., 1983, 519).

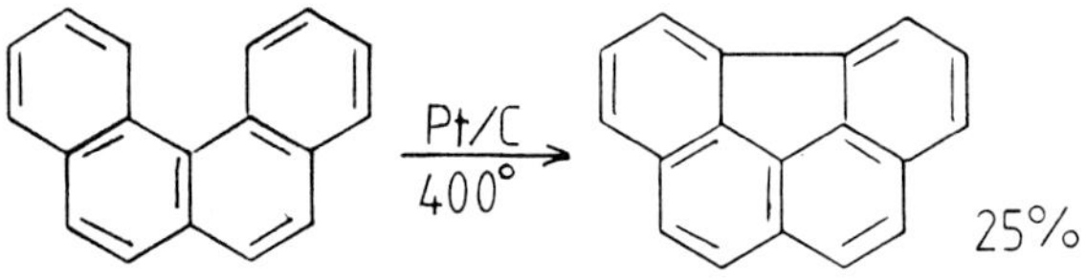

Pt/C 400° 25%

Benzo[ghi]fluoranthene

The fluorescence spectra of a wide variety of fluoranthenes have been described (S.Amin *et al.*, J.org.Chem., 1985, 50, 4642).

5 Hydrocarbons containing two five-membered rings

(a) The benzindan group

The dianions **(75)** and **(76)**, synthesised by conventional intramolecular cyclisation, provide precursors for possible ferrocenes having a helicene structure (T.J.Katz and W. Slusarek, J.Amer.chem.Soc., 1979, 101, 4259).

(75) (76)

The dianion of *s*-indacene **(77)** reacts with dichloromethylene to give, amongst other products, the anthracene valence-isomer *s*-benzvalenobenzobenzvalene **(78)**, which is remarkably stable (G.Gandillon, B.Bianco and U.Burger, Tetrahedron Letters, 1981, 22, 51).

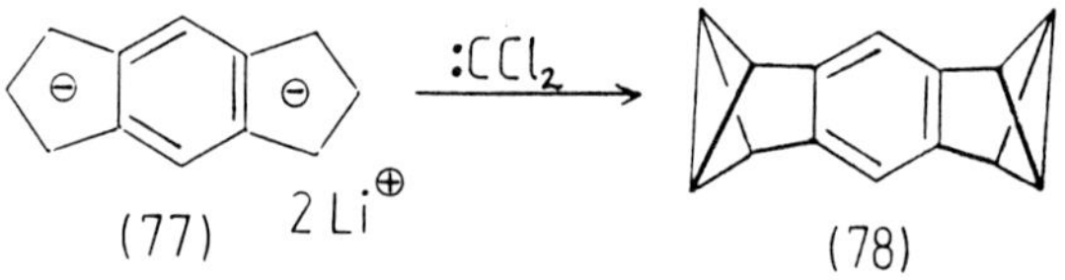

:CCl$_2$

(77) 2 Li$^{\oplus}$ (78)

Synthesis of cyclophanes derived from 1,5- and 1,7-dihydro-*s*-indacene is accomplished, not by the usual sulphur extrusion method, but by coupling di(chloromethyl)dihydroindacene using $Ni(CO)_4$ in DMF (P.Bickert, V.Boelkelheide and K.Hafner, Angew.Chem.Intern.Edn., 1982, 21, 304).

(b) The acenaphthene group

A simple synthesis of pyracylene **(79)**, previously obtainable only by a multi-step route, is effected by a flow-pyrolysis technique (G.Schaden, J.org.Chem., 1983, 48, 5385).

or

flow pyrolysis

1100°

(79)

(c) The fluorene group

When treated with SbF_5 in SO_2ClF, indeno[2,1-a]indene **(80)** affords the dication **(81)** whose proton nmr spectrum indicates a considerable degree of charge delocalisation over all four rings, (I.Willner, J.Y.Becker and M. Rabinovitz, J.Amer.chem.Soc., 1979, 101, 395).

SbF_3

SO_2ClF

(80) (81)

The corresponding dianion **(83)**, prepared by treating 5,10-dihydroindeno[2,1-a]indene **(82)** with butyllithium in THF-d_8, also shows evidence of charge delocalisation.

BuLi

(82) (83)

The dihydro compound **(82)**, which is a *trans*-stilbene analogue, undergoes photochemical [2 + 2] cycloadditions with various alkenes to provide interesting propellanes (S.C.Shim., J.S.Chae and J.H.Choi, J.org.Chem., 1983, 48, 417). An example is given below.

(82) + COOMe —hν→ COOMe

The tetrahydroderivative **(84)** of indeno[1,2-a]indene is produced when 1-phenyl-1,2-dihydronaphthalene is irradiated at 254 nm. for 20 hours, a much more direct procedure than previous methods (W. Laarhoven, F.A.T.Lijten and J.M.M.Smits, J.org.Chem., 1985, 50, 3208). X-ray analysis reveals that **(84)** has the *cis*-configuration.

H Ph —hν→ (84)

The compound **(85)**, formed by the reaction of 2,2',6,6'-tetra(hydroxymethyl)biphenyl and benzene in concentrated sulphuric acid, undergoes a remarkable rearrangement when reacted with butyllithium in HMPA, to give the dibenzopentalene derivative **(86)** (M.Rabinovitz and A.Minsky, Pure and appl.Chem., 1982, 54, 1005).

1. BuLi
2. H_2O

(85) (86)

The formation of **(86)** is a reflection of the stability of its aromatic dianion (visualised as a perturbed [12]annulene with 14 π-electrons delocalised in the periphery) which is postulated as a reaction intermediate.

(d) Other hydrocarbons containing two 5-membered rings

The Diels-Alder addition of benzyne and diazoketones to give spiroindazoles is a useful starting-point for the synthesis of a variety of hydrocarbons with fused five- and six-membered rings (2nd Edn., IIIH, p.204; W.Burgert, M.Grosse and D.Rewicki, Ber., 1982, 115, 309; H.Duerr and A.Hackenberger, Synthesis, 1978, 594).

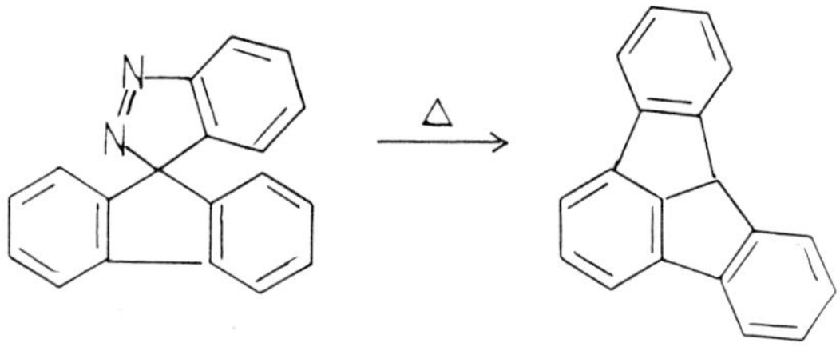

Δ

7bH-indeno[1,2,3-jk]fluorene
("Fluoradene")

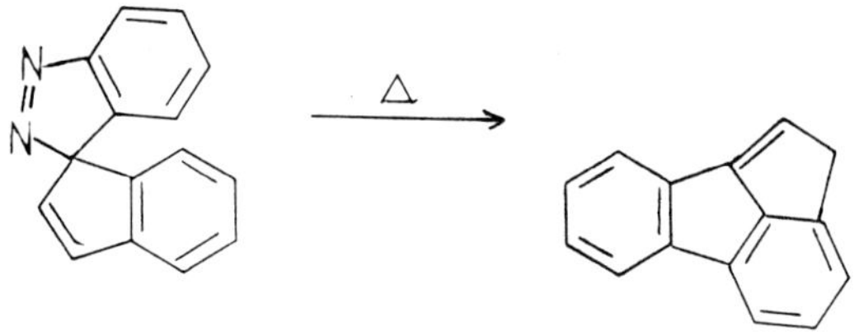

Δ

2H-cyclopenta[jk]fluorene

+

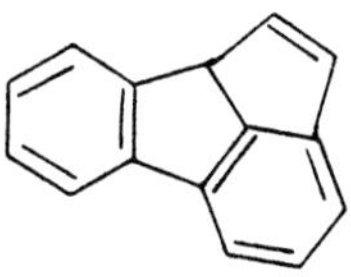

9bH-cyclopenta[jk]fluorene

Ph

Ph

hν

Ph

Ph

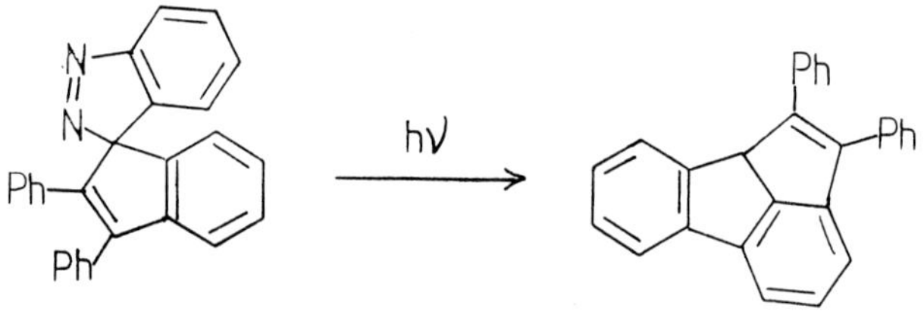

Cyclopenta[jk]fluorene, referred to above, may be regarded as a benzologue of the simpler system 1H-cyclopent[cd]indene **(87)**, which has been reported previously (K.Hafner, Angew.chem.Intern.Edn., 1971, 10, 751; B.L.McDowell and H.Rapoport, J.org.Chem., 1972, 37, 3261). A more recent synthesis is shown below (C.Wentrup and J. Benedikt, *ibid.*, 1980, 45, 1409).

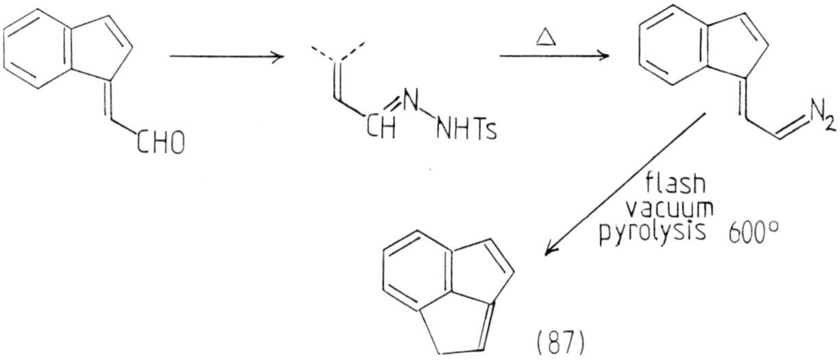

CHO

CH N NHTs

Δ

N_2

flash vacuum pyrolysis 600°

(87)

The bridged annulenes **(88)** and **(89)**, formally derived from cyclopent[cd]indene and cyclopenta[jk]fluorene respectively, are of considerable theoretical interest. Their syntheses, properties and reactions, together with those of some of their derivatives, are described in a series of papers by C.W.Rees and co-workers (C.W.Rees *et al.*, Chem. Comm., 1980, 689, 691; 1981, 657, 1065; C.W.Rees, T.L.Gilchrist and D.Tuddenham, J.chem.Soc., Perkin I, 1983, 83; R.McCague, C.J.Moody and C.W.Rees, *ibid.*, 1984, 165, 175; C.W.Rees *et al.*, *ibid.*, 1984, 909).

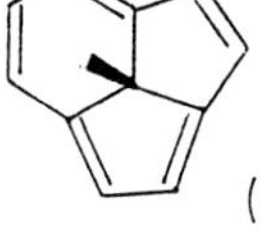

(88)

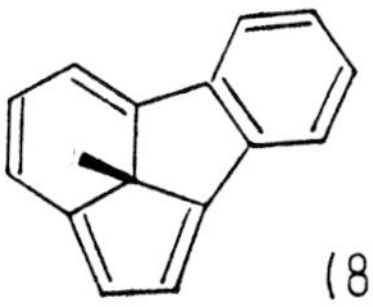

(89)

Diels-Alder cycloaddition of dimethyl acetylenedicarboxylate to 3-methoxy-3a-methyl-3aH-indene provides the required skeletal structure, and loss of methanol (by dissolution in H_2SO_4/CH_3OH) results in the formation of the diester **(90)**, from which are derived both **(88)** and **(89)**.

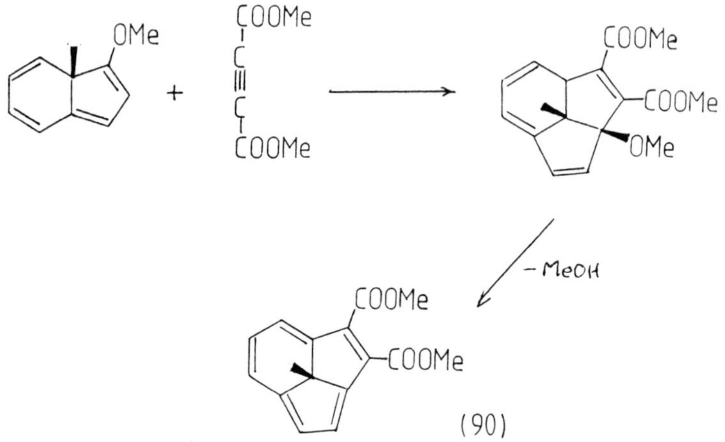

(90)

The physical properties of **(88)** and the diester **(90)** are consistent with those of a 10π aromatic periphery which sustains a diamagnetic ring current, but the fusion of a benzene ring (as in **(89)**) markedly affects the properties of the system, reducing the ring current by one third. Not unexpectedly, the protons of the central methyl group appear upfield of TMS, in the case of **(88)** at δ -1.34. The hydrocarbon **(88)** undergoes electrophilic substitution (nitration, acetylation, formylation and sulphonation) at C-5 and C-1. Cycloaddition occurs less readily, but reaction has been accomplished with certain reactive molecules such as 4-phenyl-1,2,4-triazole-3,5-dione. The ring-strain inherent in the system is probably responsible for its facile reduction (H_2/Pd/C at atmospheric pressure) to the fully saturated hydrocarbon. Although photochemically stable, **(88)** rearranges thermally by a [1,5] sigmatropic shift of the central methyl group to give the 2a-methyl isomer **(91)**, with an activation energy of 33 K cal mol^{-1}.

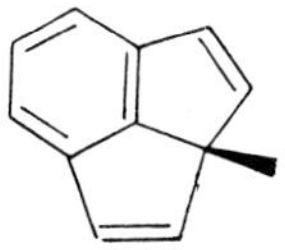

(91)

Of particular interest is the fact that in the 2-hydroxy derivative of **(88)**, the tautomeric equilibrium lies far on the side of the keto form, whereas the 5-hydroxy isomer exists entirely in the "phenol" form (C.W.Rees *et al.*, *ibid.*, 1985, 383). These results are in agreement with MNDO SCF-MO calculations (H.S.Rzepa, J.chem.Res., 1982, (S), 324; (M), 3301). Other M.O. calculations on the cyclopentindene system, referred to for this purpose as 1,4,7-methino[10]annulene, are reported by R.C.Haddon and K.Raghavachari (J.Amer.chem.Soc., 1985, 107, 289)

6 Hydrocarbons containing 5- and 7- membered rings

Cyclohept[fg]acenaphthylene **(92)** often confusingly referred to as "acepleiadylene" (cf. 2nd Edn., IIIH, p.206) attracts attention because it appears to exist as a vinyl-bridged[14]annulene, constituting an example of Platt's perimeter rule (J.chem.Phys., 1954, 22, 1448).

(92)

A dianion, and surprisingly, a tetra-anion of **(92)** are formed on reduction with lithium. In the proton nmr spectrum of **(92)**, low-field resonances point to the existence of a peripheral diamagnetic ring current, but the pronounced upfield shifts upon dianion formation (by about

8 ppm), far beyond those predicted on the basis of charge-induced shielding, are indicative of a perimeter system of 16 π-electrons (B.C.Becker, W.Huber and K.Mullen, J.Amer.chem.Soc., 1980, 102, 7803).

A 4nπ perimeter is also proposed for pyracylene **(79)**, for which the above dianion is a true homologue.

Although **(92)** is apparently inert to cycloaddition, its dihydro-derivative, cyclohept[fg]acenaphthene, does form an adduct with maleic anhydride (Y.Sugihara *et al*., Tetrahedron Letters, 1982, 23, 1925).

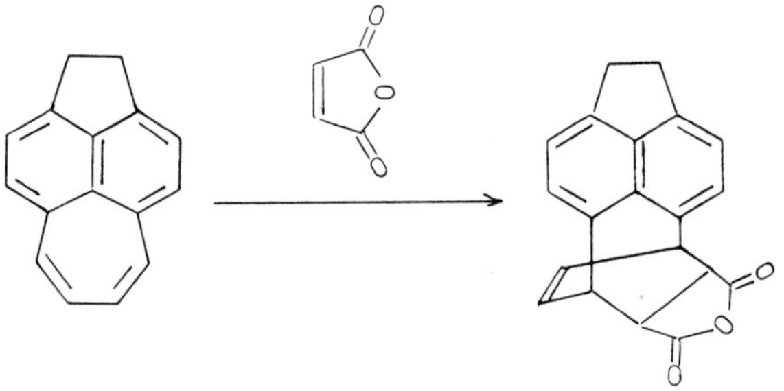

A novel route to benz[a]azulene **(95)**, involves the thermolysis of the tosylhydrazone salt **(93)** in the presence of benzene to give the dihydro-derivative **(94)**. Successive hydride abstraction and deprotonation affords **(95)**. (M.A. O'Leary and D.Wege, Tetrahedron Letters, 1978, 31, 2811). For another synthesis, see M.E.Jason, *ibid*., 1982, 23, 1635).

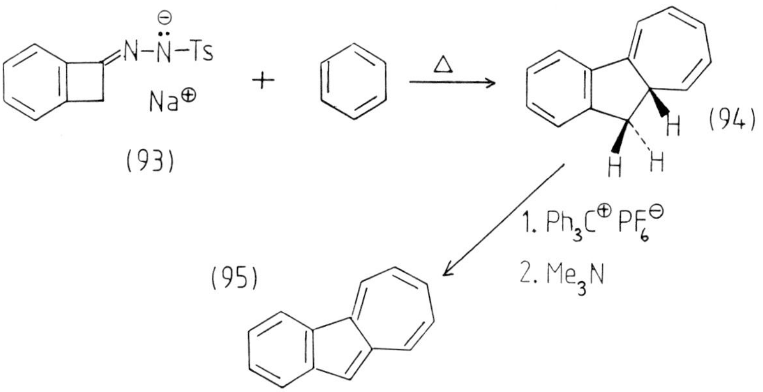

(93) + → (94)

1. $Ph_3C^{\oplus} PF_6^{\ominus}$
2. Me_3N

(95)

Interest in peripheral π-systems has prompted the synthesis of the hydrocarbon **(97)**, in analogy with **(88)** (p.67) and the elusive 9b-methylphenalene **(96)** (K.Hafner and V.Kuhn, Angew.Chem.intern.Edn., 1986, 25, 632).

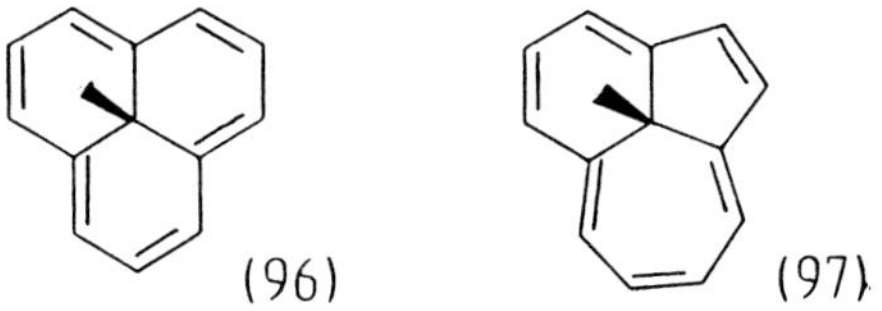

(96) (97)

The anion of 4,5-dihydro-3H-benz[cd]azulene is methylated, and the resulting compound successively brominated and dehydrobrominated to give the blue hydrocarbon **(97)**. The latter proves to be an antiaromatic [12]annulene *par excellence* among 12π systems. The perimeter protons resonate at δ3.88-4.69, and those of the central CH_3 group appear at δ4.75, suggestive of a pronounced paramagnetic ring-current.

The anion **(100)** from the hydrocarbon **(99)**, itself produced by methylation of the dianion of aceheptylene **(98)**, is a [13]annulenide ion, with a 14π perimeter. (B.C.Becker *et al.*, Angew.Chem.intern.Edn., 1983, 22, 241).

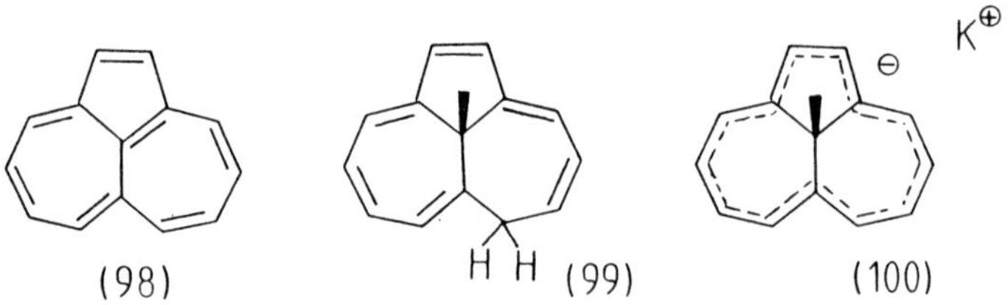

(98) (99) (100)

The proton nmr spectrum of **(100)** is notable for the extreme upfield resonance of the central methyl protons (δ -3.75), and, despite the excess charge, pronounced deshielding of the ring protons. The ion **(100)** is therefore diatropic, as predicted.

Azuleno [1,2,3-cd]phenalene **(101)**, a non-alternant isomer of benzo[a]pyrene, has proved to be as highly mutagenic as the latter (K.Nakasuji, T.Nakamura and I.Murata, Tetrahedron Letters, 1978, 18, 1539).

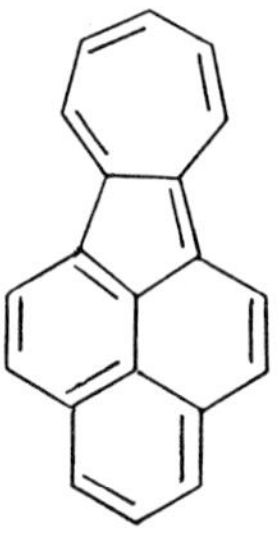

(101)

Other non-alternant isomers of benzo[a]pyrene, *viz.* benzo[4,5]cyclohept[1,2,3-bc]acenaphthylene and benzo-[a]naphth[3,4,4a,5-cde]azulene, have been synthesised in order to compare their biological activity with that of benzopyrene (E.Todo, K.Yamamoto and I.Murata, Chem.Letters, 1979, 537).

Chapter 30

POLYCYCLIC AROMATIC COMPOUNDS WITH FOUR OR MORE SIX-MEMBERED FUSED CARBOCYCLIC RING SYSTEMS

H.F. ANDREW

1 Introduction and General Properties

The definitions, general properties and nomenclature rules reviewed in the Second Edition require no elaboration in this supplement, and the opportunity has been taken to drop the various alternative names given in the former work. Except when a trivial name is in common use, only the IUPAC name is employed.

Interest in the polycyclic aromatic hydrocarbons (PCAHs) in the last decade has focussed mainly upon theoretical studies relating to aromaticity, and on the development of new, efficient syntheses linked to the study of carcinogenesis. It will be noted from the sequel that the major emphasis has been on PCAHs which have most relevance to these programmes, whereas many of the multi-ring hydrocarbons described by E.Clar ("Polycyclic Hydrocarbons", Academic Press, 1964) have received little or no attention since publication of C.C.C., 2nd Edition.

(a) General chemical and physical properties

(i) Structure-property relationships

The concept of "aromaticity" continues to fascinate theoretical chemists and provide new challenges in synthesis. (See contributed papers to the International Symposium on Aromaticity, Dubrovnik, 1979, in Pure and appl.Chem., 1980, 52, 1399-1667.) The "ring-current" concept in nmr spectroscopy is presently the method most frequently used for diagnosing and defining aromaticity (R.B.Mallion, *ibid*., p.1575).

There have been many attempts to categorise and define the topology of polycyclic aromatic hydrocarbons.

For example, dualist graph theory (A.T.Balaban, *ibid.*, 1982, 54, 1075) distinguishes "catafusenes", *e.g.* benz[a]anthracene, "perifusenes", *e.g.* pyrene and "coronafusenes", *e.g.* Kekulene (see p129), and applies topological rules to predict whether a polyhex will have a stable structure, or exist as a free-radical. In another graph-theoretical approach, a formula periodic table (J.R.Dias, Acc.chem. Res., 1985, 18, 241) is used to obtain formula-structure and physico-chemical correlations.

Historically, correlation of theoretical models for the reactivity of aromatic hydrocarbons with published experimental results has been rather unsatisfactory, due in some measure to wide variations in practical procedures. A recent systematic study of the rates of reaction of a large number of catafusenes with maleic anhydride (D.Biermann and W.Schmidt, J.Amer.chem.Soc., 1980, 102, 3163, 3173) interprets the results in terms of six different theoretical models: (a) Clar's sextet theory (2nd Edn., Vol.IIIH, p.223), (b) Brown's *para*-localisation concept, (c) Herndon's structure-count method (*loc.cit.*, p.219), (d) Free-valence indices, (e) Fukui's frontier-orbital theory and (f) complete 2nd order perturbation theory. While all six correctly account for the observed positional selectivity (nearly all reactions are regioselective), they meet with varying success with respect to rate constants. Models (c) and (f) correlate best with the observed data.

(ii) Spectroscopy

The current tendency to replace long unambiguous syntheses with shorter but less specific routes (see below) places more dependence upon spectroscopic methods for identification of the products. Nmr spectroscopy is often the method of choice in PCAH chemistry, and the publication of a general procedure for the complete assignment of proton and ^{13}C nmr spectra is most timely (D.M.Jerina *et al.*, J.org. Chem., 1985, 50, 3029). Application of the technique to the study of aromaticity has been mentioned above.

(iii) Reduction

One of the major problems in PCAH chemistry is the lack of methods for functionalisation of ring positions which are not susceptible to direct substitution. Since electrophilic, nucleophilic and radical substitution all occur regioselectively at a small fraction of total ring positions, the majority of substituted PCAHs are synthetically inaccessible except *via* tedious, individual, multi-step routes. However, the hydroaromatic compounds obtained on partial hydrogenation of PCAHs may undergo substitution at benzylic sites or at aromatic ring positions. Since these sites are likely to differ from those susceptible to substitution in the parent fully-aromatic hydrocarbon, certain other substituted derivatives are potentially available via the sequence: hydrogenation, substitution, dehydrogenation. Unfortunately, hydrogenation of PCAHs is one of the least predictable and controllable reactions of these compounds (2nd Edn, Vol.IIIH, p.224). A major contribution to the solution of this problem has been provided by P.P.Fu, H-M.Lee and R.G.Harvey (J.org.Chem., 1980, 45, 2797). In several PCAHs studied, hydrogenation over palladium at low pressure and ambient temperature gives regiospecifically the corresponding K-region* dihydro derivatives, while the analogous reactions over platinum occur regiospecifically in terminal rings to give tetrahydro compounds as illustrated below.

$H_2/Pd/C$

H_2/Pt

* A K-region is a bond, such as the 9,10-bond of phenanthrene, excision of which from a polynuclear hydrocarbon leaves an intact aromatic ring system.

(b) *Carcinogenesis*

The topic of carcinogenicity of polycyclic aromatic hydrocarbons (PCAHs) is too wide to be discussed here in detail, but since the major motivation for much of the chemistry described in this chapter derives from attempts to understand the mechanism of mutagenicity, a short summary of the current (1987) state of knowledge may be helpful: Carcinogenic PCAHs require activation by the P-450 microsomal enzymes in order to exhibit their mutagenic and carcinogenic potential. For example, for benzo[a]pyrene, the most intensively investigated hydrocarbon, the principal biologically-active metabolite has been identified as (+)*trans*-7,8-dihydroxy-*anti*-9,10-epoxy-7,8,9,10-tetrahydrobenzo[a]pyrene, (*anti*-BPDE). In mammalian cells, covalent binding of *anti*-BPDE to DNA and RNA takes place principally on guanosine, yielding a 2-aminodeoxyguanosine adduct. The mechanism appears to involve rapid initial formation of an *anti*-BPDE—DNA intercalation complex, followed by a slow protonation to yield an intercalated triol carbocation intermediate that undergoes either hydration to tetrols (major path) or covalent binding to DNA (minor path), and rearrangement of this adduct to a more stable structure in which the PCAH (benzopyrene) ring system is bound externally in the minor groove of the DNA helix (J. Pataki and R.G.Harvey, J.org.Chem., 1987, 52, 2226). A fuller review of this topic is given by R.G.Harvey in Synthesis, 1986, 605.

(c) *General methods of preparation*

Two developments in the synthesis of PCAHs can be discerned. One, which has been mentioned above, is a trend to shorter, more direct routes. Another is the improvement of certain key steps by the application of more selective reagents. This is well illustrated by dehydrogenation, which is frequently the last step in the synthesis of PCAHs and their derivatives. The older methods for this reaction tend to employ sulphur, selenium or platinum under drastic conditions unsuitable for the preparation of the more sensitive compounds currently of greatest interest. An additional limitation is the impracticability of controlling the extent of the reaction to obtain partially-aromatised compounds which themselves are valuable intermediates. Newer, milder reagents, less hampered by these limitations, are now supplanting the older ones. Among these may be mentioned the high oxidation-potential quinones (DDQ, chloranil), trityl salts, alkyllithium-amine complexes and *N*-bromosuccinimide. A comprehensive review of the methods is available (P.P.Fu and R.G.Harvey, Chem.Rev., 1978, 78, 317).

Four synthetic strategies, examples of which are distributed throughout the sequel, have found extensive application.

(i) The use of benzyne, naphthyne and naphthodiynes as dienophiles in cycloaddition reactions. Such reactive intermediates, formed *in situ* from suitable precursors, are allowed to react with heterocyclic dienes; subsequent elimination of the cheletropic bridge from the adduct furnishes a new benzene ring.

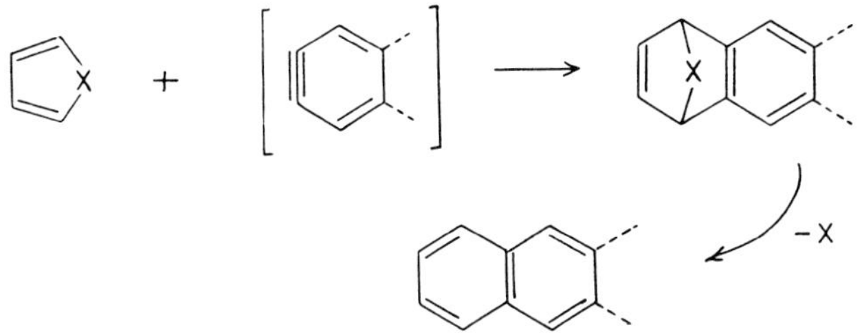

(ii) Prior elimination of a cheletropic bridge from a precursor molecule to produce a transient xylylene which can undergo cycloaddition to a conjugated double-bond.

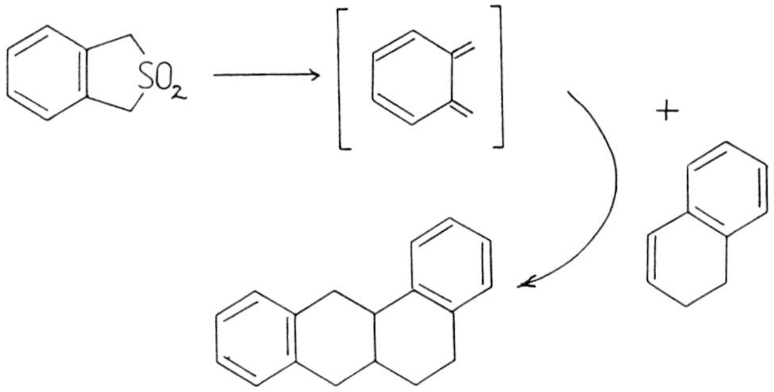

(iii) Cyclodehydration of a divinylalcohol formed from an *ortho*-quinone.

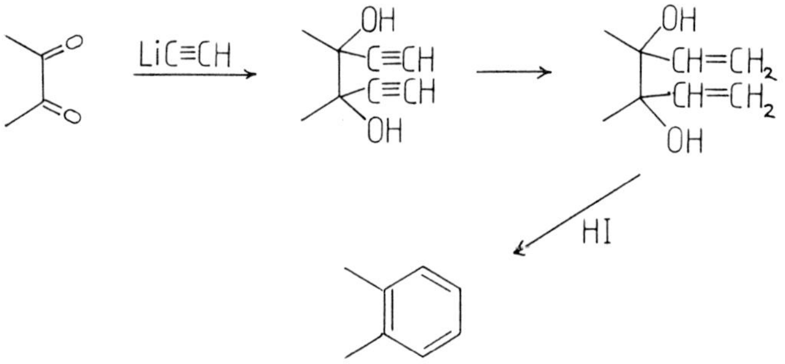

LiC≡CH

OH
C≡CH
C≡CH
OH

OH
CH=CH$_2$
CH=CH$_2$
OH

HI

(iv) Reaction of an aromatic carbonyl compound and an *o*-lithioarylamide to give a lactone which can undergo intramolecular cyclisation.

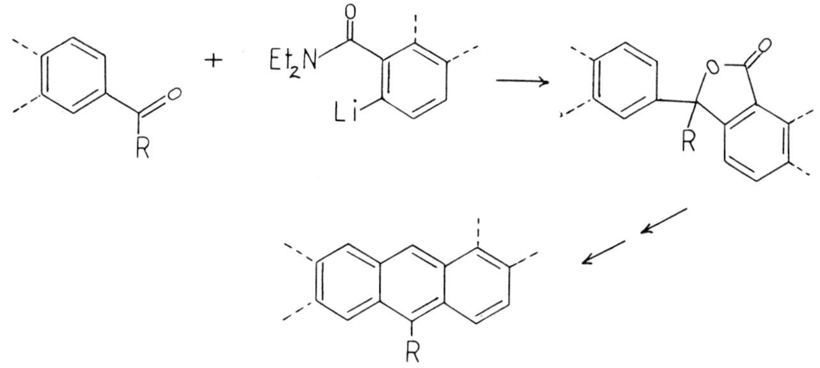

2 Individual Hydrocarbons containing Four Benzene Rings

(a) *Chrysene and its derivatives*

(i) *Synthesis*

The photochemical route from styrylnaphthalenes (2nd Edn. III H, 239) continues to find useful applications (P.H. Gore and F.S.Kamouna, Synthesis, 1978 773); F.S.Kamouna and I.S. Al-Shiebani, Chem.Abs., 1986, 105, 97139; S.Amin, S.S.Hecht and D.Hoffmann, J.org.Chem., 1981, 46, 2394). A noteworthy improvement to the method, which often suffers from low yields, is the use of an ester derivative which cyclises cleanly in 60% yield.

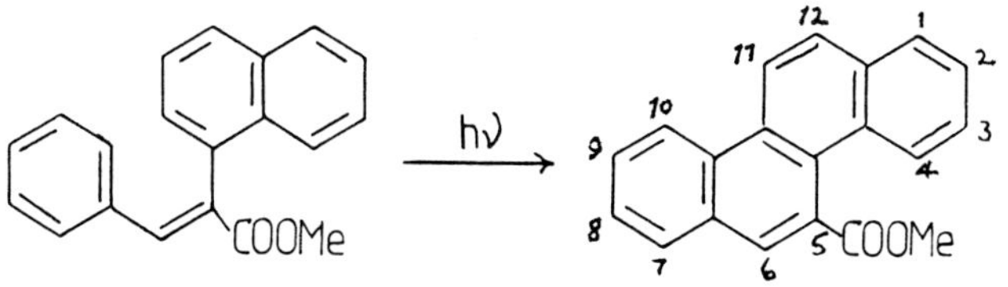

The ester function may be subsequently modified, giving rise to 5-substituted chrysenes including the highly carcinogenic 5-methylchrysene (S.Amin *et al.*, *ibid.*, 1984, 49, 381).

A new, efficient synthesis of chrysene and its derivatives features as the key step the cycloaddition between 1,5-naphthodiyne and a heterocyclic diene such as furan.

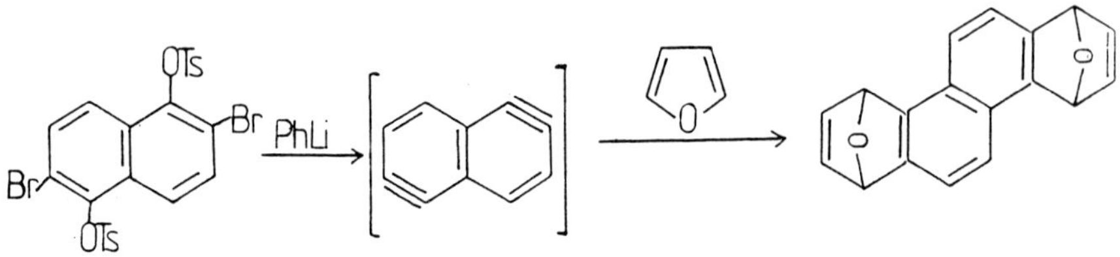

OTs
Br
PhLi
Br
OTs
O
O
O

The *bis*-adduct undergoes acid-catalysed aromatisation to chrysene (C.S.Le Houllier and G.W.Gribble, *ibid.*, 1983, 48, 1682). *N*-Methylpyrrole is an alternative addend.

Another generally applicable synthesis is based upon the intramolecular capture of an *o*-quinodimethane (xylylene) generated by cheletropic elimination of SO_2 from a strategically-designed and easily-accessible precursor (L.A.Levy and V.P.Shashikumar, *ibid.*, 1985, 50, 1760).

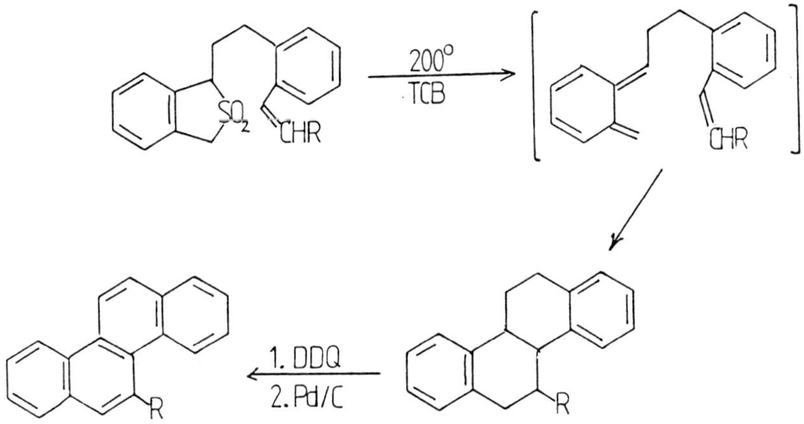

SO_2
CHR
200°
TCB
CHR
1. DDQ
2. Pd/C
R
R

A method which is particularly useful for the synthesis of certain hydroxychrysenes required for carcinogenic studies, and which is claimed to be superior to the photochemical route (S.Amin, *loc.cit*.), is outlined below (R.G.Harvey, J.Pataki and H.Lee, J.org.Chem., 1986, 51, 1407).

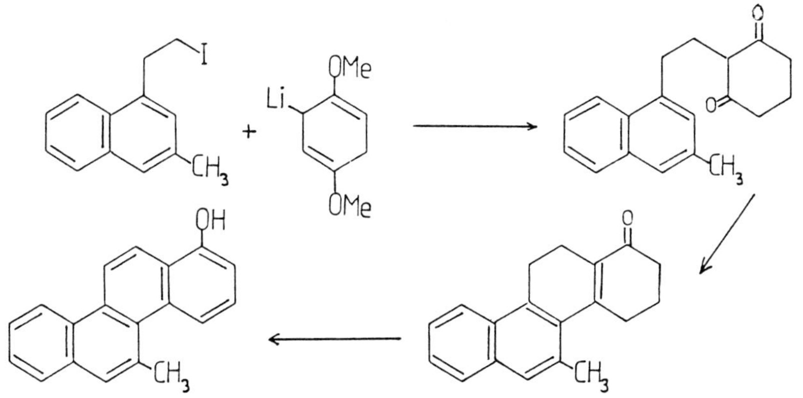

5,6,11,12 - Tetrahydrochrysene and its derivatives may be obtained *via* the reaction of 1-tetralones with 2-bromoethylbenzenes (T.A.Lyle and G.H.Daub, *ibid*., 1979, 44, 4933).

(ii) Reactions

Chrysene is nitrated conveniently and quantitatively with N_2O_4 in dichloromethane, the product containing over 90% 6-nitrochrysene (F.Radner, Acta.chem.Scand., 1983, 37B, 65).

The difficulty in obtaining *t*-butylchrysenes has been solved by an elegant conversion of 6-acetylchrysene into 6-*t*-butylchrysene (J.Pataki, M.Konieczy and R.G.Harvey, J.org.Chem., 1982, 47, 1133).

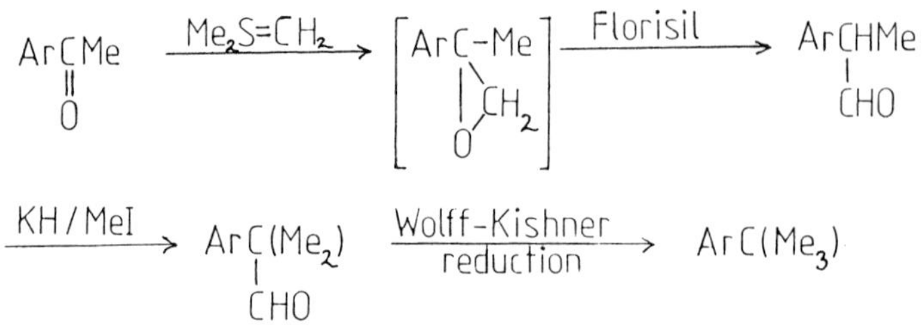

$$ArC(=O)Me \xrightarrow{Me_2S=CH_2} [ArC(Me)(CH_2O)] \xrightarrow{Florisil} ArCH(Me)CHO \xrightarrow{KH/MeI} ArC(Me_2)CHO \xrightarrow[\text{reduction}]{\text{Wolff-Kishner}} ArC(Me_3)$$

As mentioned in the introduction, diols, epoxides and diol-epoxides are of current interest in the study of mutagenicity. Those of chrysene are described by P.P.Fu and R.G. Harvey, Chem.Comm., 1978, 585; D.M.Jerina *et al*., J.org. Chem., 1982, 47, 1110; D.R.Boyd, M.G.Burnett and R.M.E. Greene, J.chem.Soc., Perkin I, 1983, 595; P.J.Van Bladeren and D.M.Jerina, Tetrahedron Letters, 1983, 24, 4903.

The structure of 5-methylchrysene has been determined. The crystal is monoclinic, space group $P2_1/C$. One of the molecules is disordered due to repulsions between the methyl group and the hydrogen at C-4, resulting in out-of-plane displacement of the methyl carbon and increased bond angles (122°-127°) in the bay region between C-4 and C-5 (R.G. Harvey *et al*., Acta Cryst., 1984, 40C, 536).

(b) Naphthacene and its derivatives

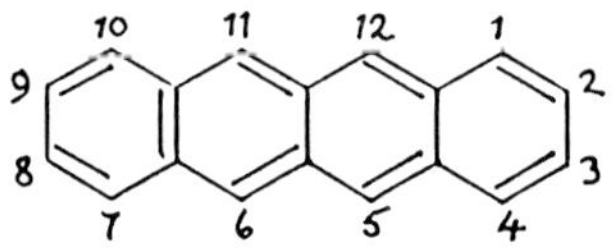

Naphthacene is non-carcinogenic, but the ring-system attracts attention as the basis of tetracycline and anthracyclinone antibiotics (2nd Edition IIIH, 249; M.J. Morris and J.R.Brown, Tetrahedron Letters, 1978, 32, 2937).

(i) Synthesis

The older methods outlined in the 2nd Edition are mostly unsatisfactory, necessitating the development of new routes. Sequential aryne cycloaddition and cheletropic bridge extrusion, already described in connection with chrysene (p.78), is further exemplified in the following naphthacene syntheses (C.S.Le Houllier and G.W.Gribble, J.org.Chem., 1983, 48, 2364; G.W.Gribble, R.B.Perni and K.D.Onan, *ibid*., 1985, 50, 2934).

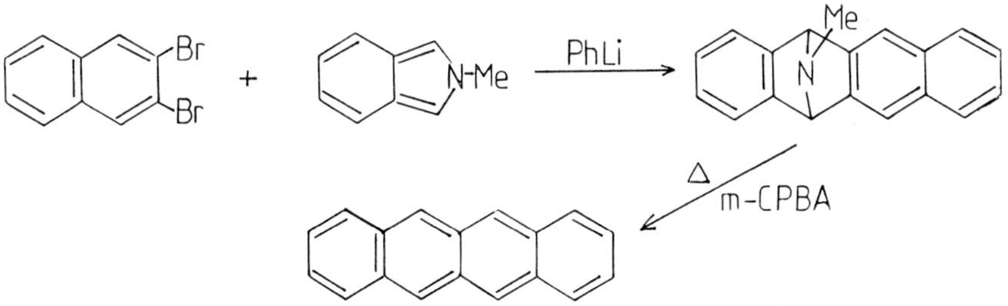

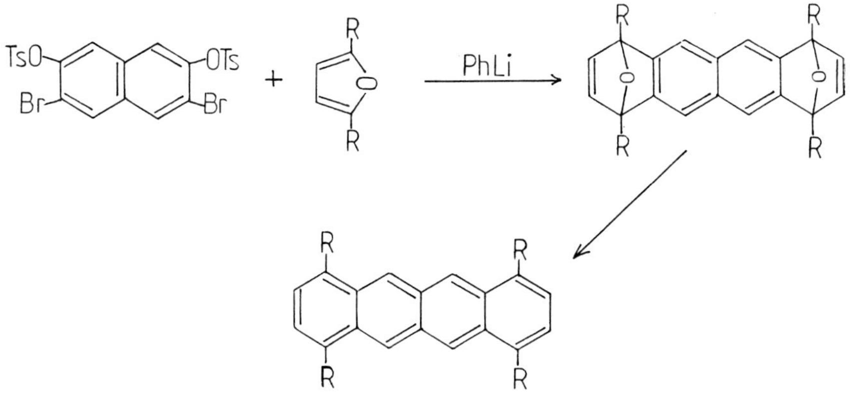

Br Br + N-Me —PhLi→ (N-Me bridged adduct) —Δ, m-CPBA→ naphthacene

TsO OTs Br Br + (2,5-R-furan) —PhLi→ (bis-oxa-bridged adduct, R R R R) → 1,4,7,10-tetra-R-naphthacene

Naphthacene-2,3-dialdehyde is formed when benzene-1,2-dialdehyde is condensed with three equivalents of butane-1,4-dial (formed *in situ* from 2,5-dimethoxytetrahydrofuran). Two equivalents and one equivalent give respectively anthracene dialdehyde and naphthalene dialdehyde (A.Mallouli and Y.Lepage, Synthesis, 1980, 689).

Another preparation, of mechanistic interest rather than synthetic utility, is that of 5,12-dihydronaphthacene (5) from diethynylbenzene (1) and the ditosylate (2).

The intermediates (3) and (4) are postulated (H.N.C.Wong and F.Sondheimer, Tetrahedron Letters, 1980, 21, 217).

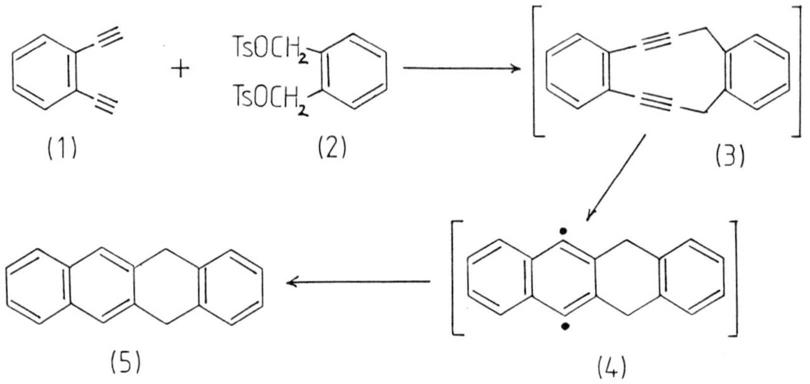

TsOCH$_2$
TsOCH$_2$

(1) (2) (3)

(5) (4)

Permethylnaphthacene, $C_{30}H_{36}$, reddish-orange crystals, m.p. 265-266° is of interest because of its twelve "peri" interactions between CH_3 groups. (A.Sy and H.Hart, J.org. Chem., 1979, 44, 7). The proton nmr spectrum shows 3 equal singlets, δ 2.98, 2.66 and 2.32, which can be assigned to the three sets of CH_3 groups (proceeding from the centre to the ends of the molecule), and the CH_3 carbons appear in the ^{13}C spectrum at δ 28.55, 22.57 and 16.32. Permethylnaphthacene is synthesised in three steps from tetramethylnaphthalene by means of the aryne cycloaddition-cheletropic extrusion route referred to above.

The total synthesis of pretetramid (6), a precursor for certain tetracyclines, is accomplished by annelation of an appropriately-substituted anthracene. (J.A.Murphy and J. Staunton, Chem.Comm., 1979, 1166).

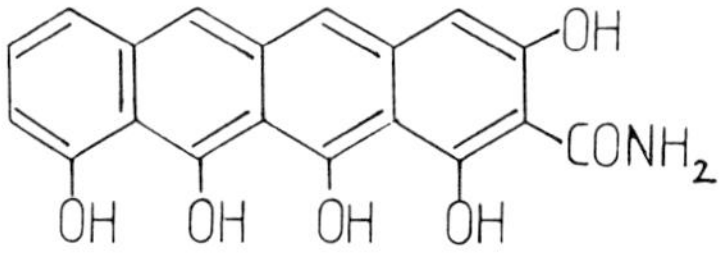

Pretetramid (6)

Many other compounds based upon the naphthacene nucleus have been synthesised in connection with antibiotics research, but these lie outside the scope of this review (A short introduction is given in the 2nd Edition, IIIH, p.249).

(ii) Reactions

When 5,12-dihydronaphthacene (5) is dissolved in high purity, aprotic molten antimony trichloride containing 10% aluminium chloride at 100-130°, the naphthacenium ion (8) is formed (A.C.Buchanan, A.S.Dworkin and G.P.Smith, J.org.Chem., 1981, 46, 471)

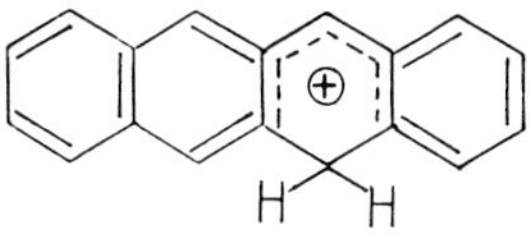

(8)

A prominent feature in the proton nmr spectrum of (8) is the presence of a midfield peak for the hydrogens on the sp^{3}-hybridised carbon at δ 4.9 with an integral twice that of the low-field-shifted resonance of H-12 at δ 9.6.

The photolysis of naphthacene in degassed fluid solution has been re-examined, with the observation that two photo-dimers are formed, a centrosymmetric isomer **(9)** and a plano-symmetric isomer **(10)** (R.Lapouyade, A.Nourmamode and H.Bouas-Laurent, Tetrahedron, 1980, 36, 2311).

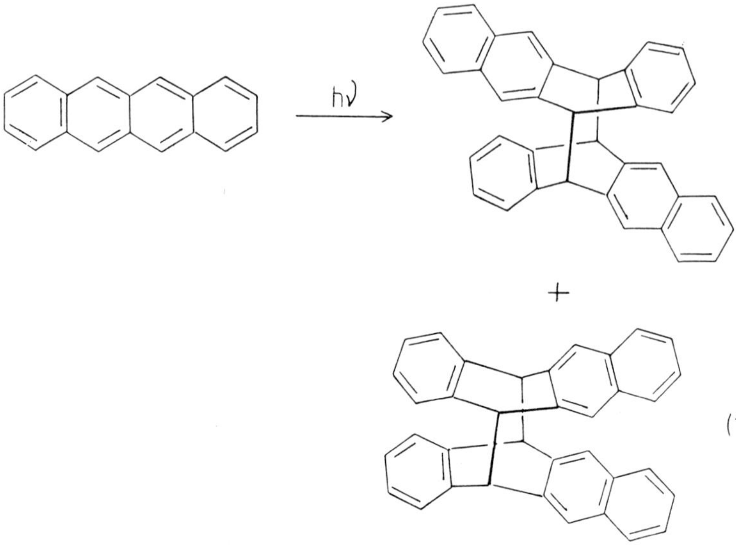

hν

(9)

+

(10)

The influence of *meso* phenyl substituents on the thermal behaviour of naphthacenic endoperoxides such as **(11)** is discussed by J.Rigaudy and D.Sparfel (Bull.Soc.chim.France, 1977, 742; Tetrahedron Letters, 1978, 34, 2263).

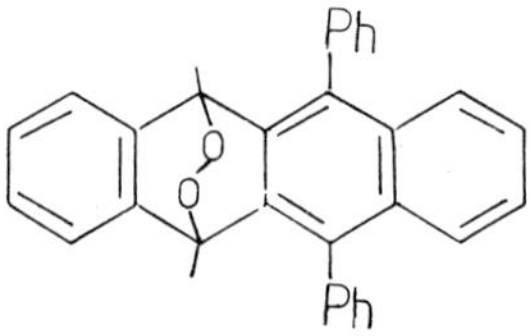

Ph
O
O
Ph
(11)

(c) *Benz[a]anthracene*

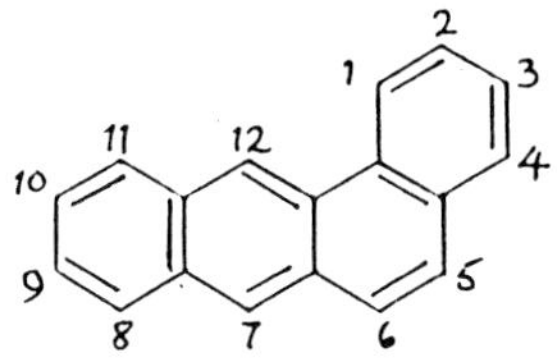

(i) *Synthesis*

Benz[a]anthracene, and more particularly certain of its methyl homologues, are carcinogenic, and consequently a great many synthetic routes to it have been devised. A review of synthetic methods for benzanthracenes (to 1977) is available (T.Otsuki and K.Maruyama, Yuki Gosei Kagaku Kyokaishi 1978, 36, 206).

The adaptability of the aryne cycloaddition-cheletropic bridge extrusion methodology is illustrated in the following two syntheses (G.W.Gribble, R.B.Perni and K.D.Onan, J.org. Chem., 1985, 50, 2934; G.W.Gribble *et al.*, *ibid.*, 1985, 50, 1611).

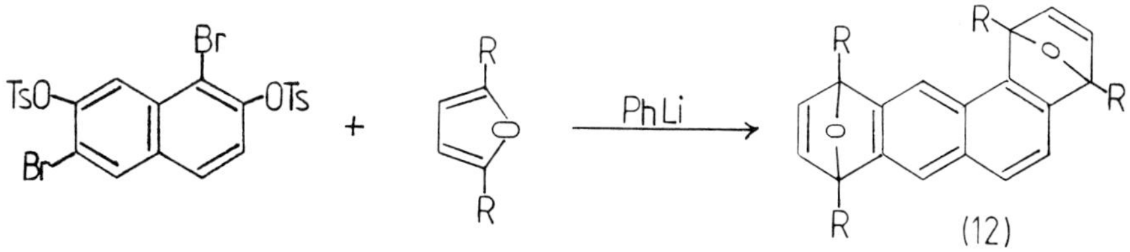

(12)

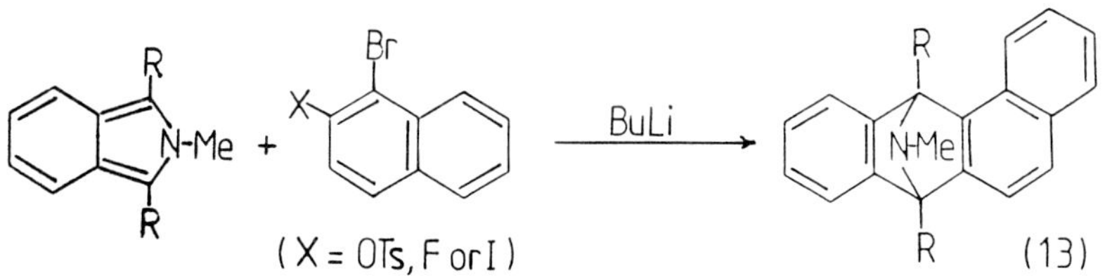

(X = OTs, F or I)

(13)

The adducts (**12**) and (**13**) are readily converted into benz[a]anthracene derivatives with substituents in the terminal and central rings respectively. Both methods can be used for benz[a]anthracene itself.

Another cheletropic elimination, in this case from 1,3-dihydrobenzo[c]thiophen-2,2-dioxide (**14**), produces the transient *o*-quinodimethane (**15**) which reacts with 1,2-dihydro naphthalenes (**16**) give the adducts (**17**) in 92% yield. Aromatisation is accomplished by treatment with DDQ to give benz[a]anthracenes (**18**) (L.A.Levy and L.Pruitt, Chem.Comm., 1980, 227).

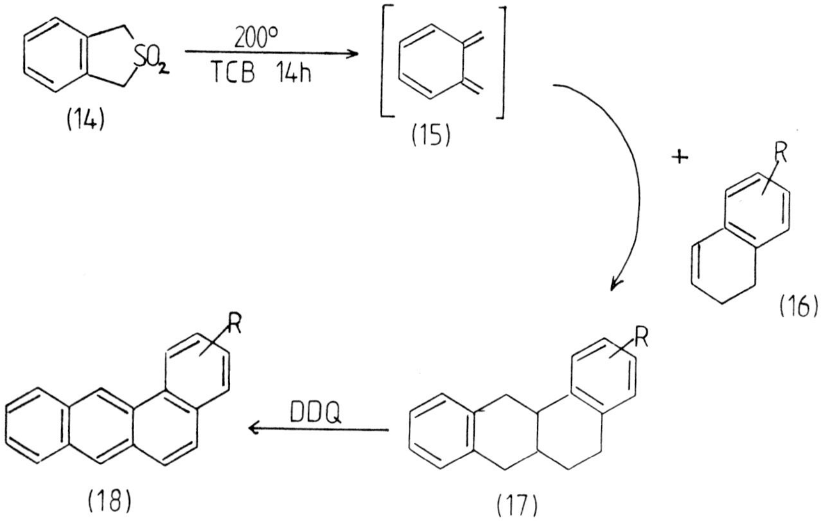

SO_2 200° TCB 14h

(14) (15) (16) (17) (18)

R

DDQ

A useful synthon for this type of synthesis is 1,3-dimethylisobenzofuran (**19**). With cinnamic esters it affords adducts (**20**) which are readily converted into a variety of 7,12-dimethylbenz[a]anthracene derivatives (**21**), including the sterically congested 1,7,12-trimethyl derivative (L.A.Levy and S.Kumar, Tetrahedron Letters 1983, 24, 1221).

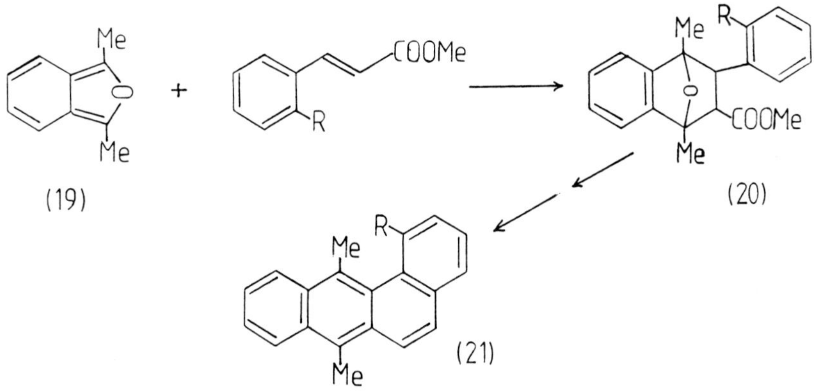

(19)

(20)

(21)

In a closely related synthesis, 1-benzylisobenzofuran and dimethyl acetylenedicarboxylate (DMAD) react to give the adduct **(22)**, which leads to benzanthracene derivatives *via* the sequence shown (J.G.Smith *et al*., J.org.Chem., 1981, 46, 4658).

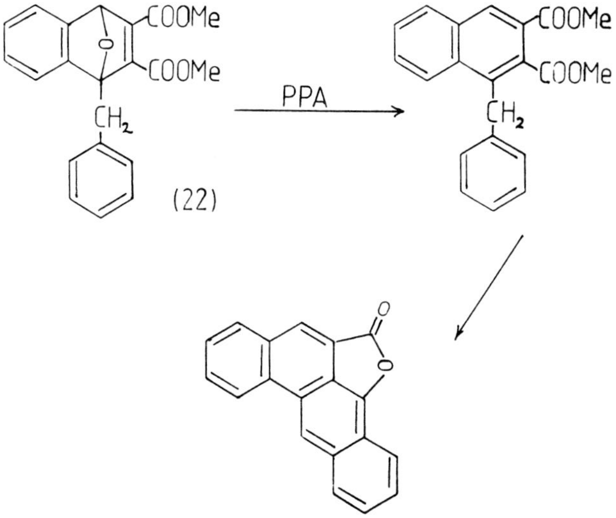

PPA

(22)

Methyl acrylate, when used as the dienophile in the above sequence, allows the facile synthesis of 7-hydroxy-benz[a]anthracenes.

Many other syntheses incorporating Diels-Alder cyclo-additions have been reported (S.W.Wunderly and W.P.Weber, J.org Chem., 1978, 43, 2277; W.Tochtermann, A.Malchow and H.Timm, Ber., 1978, 111, 1233; G.M.Muschick *et al.*, J.org. Chem., 1979, 44, 2150; K.Maruyama, T. Otsuki and K. Mitsui, *ibid.*, 1980, 45, 1424; K.Maruyama, S.Tai and T.Otsuki, Chem.Abs., 1984, 100, 191499).

R.G.Harvey's *o*-lithioarylamide route (Chapter 29, p.35) provides a useful preparation of 7-alkylbenz[a]anthracenes (R.G. Harvey, C.Cortez and S.A.Jacobs, J.org.Chem., 1982, 47, 2120).

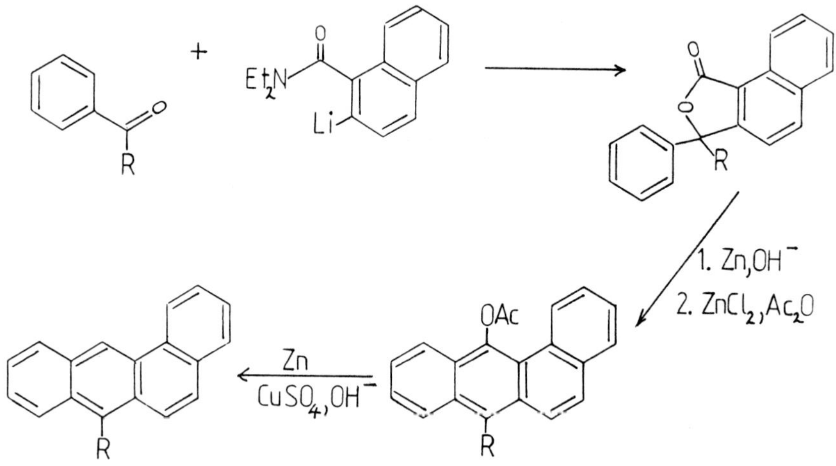

Older syntheses still find important applications, and many have been improved. For example, the reagent hydriodic acid/phosphorus has proved particularly effective in two similar roles (K.L.Platt and F.Oesch, J.org.Chem, 1981, 46, 2601; M.Konieczny and R.G.Harvey, Org.Synth., 1984, 62, 165).

The latter reaction employs less drastic conditions than previous methods, is superior to a wide range of alternatives and has wide applicability.

Synthesis of benzanthracenes by phthaloylation of naphthalene derivatives, *via* intermediates such as **(23)** and **(24)**, may be usefully modified by prior hydrogenation of one of the naphthalene rings. Such an approach allows the synthesis of 5- and 6-substituted benz[a]anthracenes (P.P.Fu *et al.*, Org.prep.Proced.Intern. 1982, 14, 169).

The task of selecting an appropriate route from such a bewildering array may be facilitated by a review of "recommended syntheses" (M.S.Newman, J.M.Khanna and K.C.Lilje, *ibid.*, 1979, 11, 271). Newman and his collaborators have applied the recommended routes to the preparation of several substituted benz[a]anthracenes (M.S.Newman and co-workers, J.org.Chem., 1979, 44, 866; 1982, 47, 2837; 1983, 48, 2926; 1983, 48, 3246 and 3249; 1986, 51, 1631).

(ii) Reactions

Anodic fluorination of benz[a]anthracene, to give the 7-F, 12-F and 7,12-diF derivatives, is achieved by controlled-potential electrolysis at Ag/Ag^{+} in CH_3CN containing $(CH_3)_4$ NF.2HF (R.F.O'Malley *et al.*, *ibid.*, 1981, 46, 2816).

5-Hydroxybenz[a]anthracene is autoxidised by air in anhydrous acetone to the 5,6-dione (T.C.Yaung and W.M.Trie, Org.Prep.Proced. Intern., 1982, 14, 202). Fremy's salt $(KSO_3)_2$ NO is an alternative oxidant, and this is also effective in oxidising the 3-hydroxyisomer to the corresponding 3,4-dione, thus providing a convenient route to terminal-ring dihydrodiols (K.B. Sukumaran and R.G. Harvey, J.org.Chem., 1980, 45, 4407).

(iii) Alkylbenz[a]anthracenes

(For syntheses, see above and 2nd Edition III H, p.254). A noteworthy phenomenon, which has occasioned much speculation and research, is the extraordinarily enhanced carcinogenicity of 7,12-dimethylbenz[a]anthracene over that of the 7- and the 12-monomethyl derivative. Newman has observed that these are, respectively, a non-planar molecule of very high activity, a planar molecule of high activity, and a non-planar molecule of low activity. On the principle that an 11-methyl group should buttress the 12-group, hence increasing the strain and perhaps the carcinogenicity, he has prepared 7,11,12-trimethylbenzanthracene (M.S.Newman, J.org.Chem., 1983, 48, 3249). Initial results appear to support the prediction. On the other hand, 1,7,12-trimethylbenz[a]anthracene is completely inactive, presumably because the $1-CH_3$ interferes with the formation of the 1,2-epoxide metabolite.

(iv) Metabolites

As with all carcinogenic polycylic hydrocarbons, the oxygenated metabolites of benz[a]anthracene are a major focus of attention. The preparation and reactions of various epoxides, diols and diol-epoxides are described by M. S.Newman *et al.*, *ibid.*, 1979, 44 4994; K.Sukumarand and R.G.Harvey, J.Amer.chem.Soc., 1979, 101, 1353; G.H.Posner and J.R.Lever, J.org.Chem., 1984, 49, 2029; D.M.Jerina *et al.*, *ibid.*, 1982, 47, 1110; H.Lee and Harvey, Tetrahedron Letters, 1981, 22, 1657; D.R.Boyd *et al.*, J.chem. Soc., Perkin I, 1981, 2233.

(v) Hydrobenz[a]anthracenes

Benz[a]anthracene can be regioselectively hydrogenated (see p.73 and 2nd Edition, IIIH, p.225; P.P.Fu, H.M.Lee and R.G.Harvey, J.org.Chem., 1980, 45, 2797), and the resulting partially-hydrogenated compounds employed to introduce functional groups into ring positions not prone to direct substitution (Harvey *et al.*, *ibid.*, 1979, 44, 4265). An illustrative example is given below.

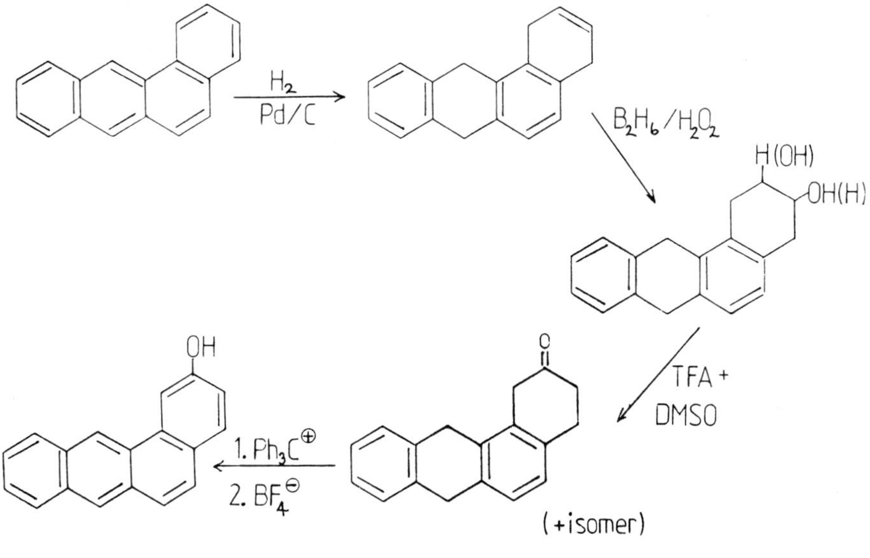

H_2, Pd/C; B_2H_6/H_2O_2; H(OH), OH(H); TFA + DMSO; (+isomer); 1. $Ph_3C^{\oplus}$ 2. $BF_4^{\ominus}$; OH

(d) Benzanthrene

7H-Benz[de]anthracene, benzanthrene **(25)** forms an anion (26) in the presence of strong base. When it reacts with *n*-butyllithium and dichloromethane, a mixture of three products **(27)**, **(28)**, **(29)** is obtained, clearly as a result of attack by chlorocarbene (R.M.Pagni, M.Burnett and A.C. Hazell, J.org.Chem., 1978, 43, 2750).

n-BuLi

(25)

(26)

BuLi + CH_2Cl_2

10%

23%

trace

(27)

(28)

(29)

The identity of the major product, the phenanthro-bicyclobutane **(28)** is established by its nmr spectrum, which is characteristic of a phenanthrene derivative, and more definitively by X-ray crystallography. Although the minor isolated product, 4,5-benzocyclohepta[1,2,3-de]naphthalene **(27)**, is obtained in only 10% yield, this short preparation is more attractive than alternative multi-step syntheses.

Benzanthrone **(30)** has for many years been the source of violanthrene vat-dyestuffs (2nd Edition, IIIH, p.261). Two new observations have been made by J.Aoki, M.Takekawa and S.Fujisawa (J.org.Chem., 1981, 46, 3922).

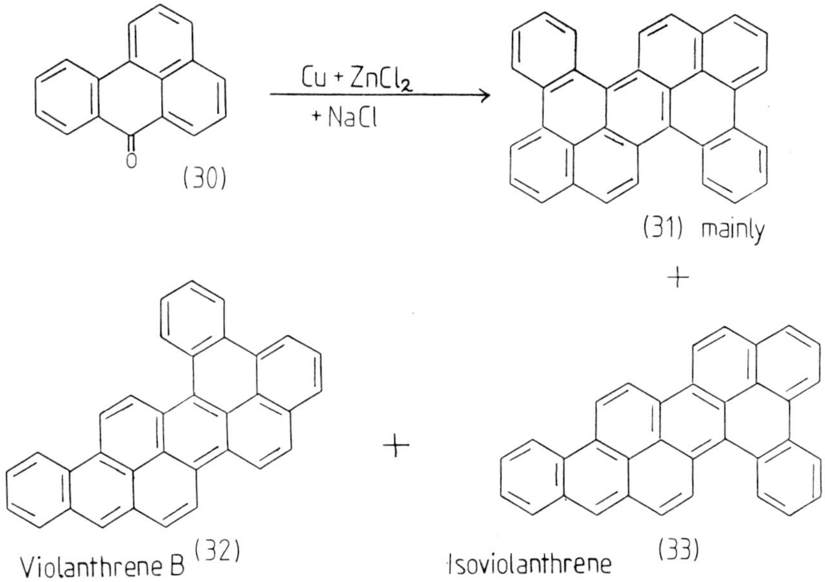

(30) $\xrightarrow[+\ NaCl]{Cu\ +\ ZnCl_2}$ (31) mainly + Violanthrene B (32) + Isoviolanthrene (33)

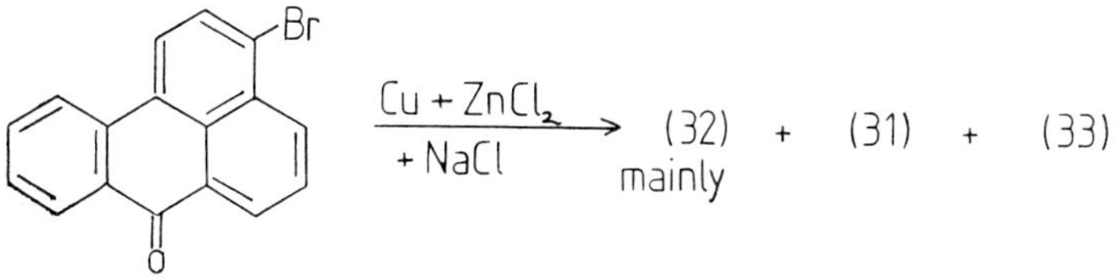

$\xrightarrow[+\ NaCl]{Cu\ +\ ZnCl_2}$ (32) mainly + (31) + (33)

The latter reaction provides a new and convenient synthesis of violanthrene B, (38% yield), previously obtained by reduction of a by-product in the conventional violanthrone synthesis.

On reaction with zinc dust, benzanthrone gives fluoranthene (see Chapter 29, p.54).

(e) *Benzo[c]phenanthrene*

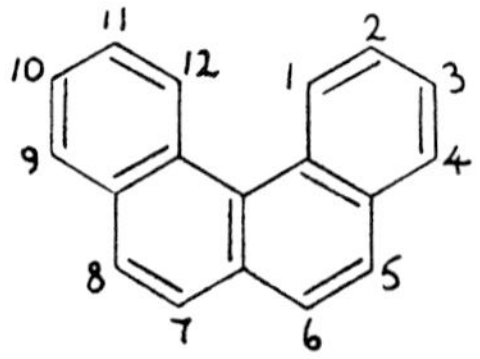

(34)

This hydrocarbon, unlike most other alternate polycyclic aromatic hydrocarbons which are carcinogenic, has a "fjord region" rather than a bay region (Chapter 29, p.38), between C-1 and C-12, and is non-planar due to congestion in this area. Two syntheses aimed towards the dihydrodiol and epoxide derivatives employ literature methods starting from naphthalene (M.Croisy-Delcey, Y.Itta and D.M.Jerina, Tetrahedron Letters, 1979, 31, 2849; D.R.Boyd *et al.*, J.chem. Soc., Perkin I, 1984, 1781).

A novel route, particularly valuable for certain hydroxy-derivatives, uses fewer steps than the foregoing methods (J.Pataki and R.G.Harvey, J.org.Chem., 1982, 47, 20).

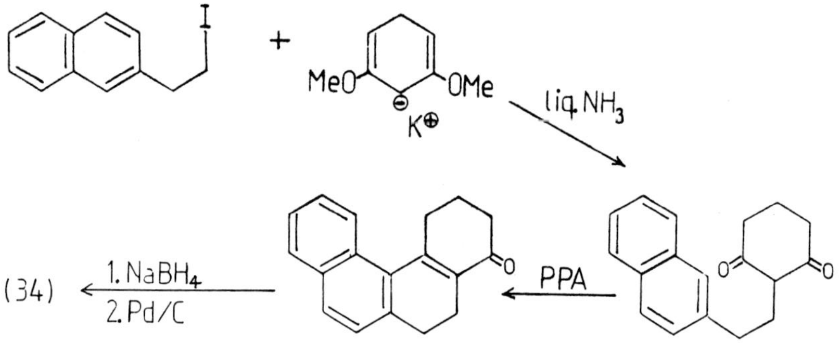

MeO OMe K⊕ liq.NH$_3$

(34) ← 1. NaBH$_4$ 2. Pd/C ← PPA

Another short synthesis, involving a double cycloaddition of maleic anhydride to 1-phenyl-1,4-dihydronaphthalene, is given by E.H.Vickery and E.J.Eisenbraun (Org.prep.Proced. Intern., 1979, 11, 261).

A facile preparation of benzo[c]phenanthrene-5,6-epoxide derives from the fact that polycyclic aromatic hydrocarbons react at the K-region (see p.73) with *N*-bromoacetamide in acetic acid to give *trans*-bromohydrin acetates, which are transformed into the epoxides on treatment with sodium methoxide in THF (P.J.Van Bladeren and D.M.Jerina, Tetrahedron Letters, 1983, 24, 4903).

(f) *Pyrene and its derivatives*

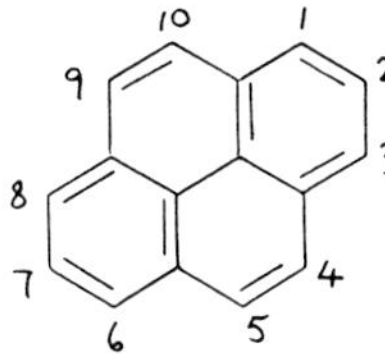

(i) *Synthesis*

In contrast to other hydrocarbons described in this section, pyrene has received more attention to its reactions than to its synthesis, a fact which may be attributable to its non-carcinogenicity and the adaptability of known synthetic routes. A useful modification of such a route is the application of methanesulphonic acid, a mild cyclising agent as an alternative to HF, PPA and sulphuric acid (A.A.Leon, G.Daub and I.R.Silverman, J.org.Chem., 1984, 49, 4544).

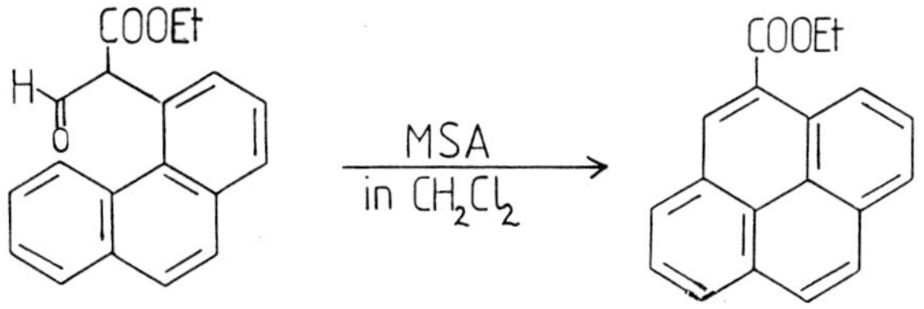

COOEt H O —MSA in CH_2Cl_2→ COOEt

Another known method, irradiation of divinylbiphenyls, has been modified in order to synthesise 4,9-difluoropyrene (R.Lapouyade, N.Hanafi and J-P.Morand, Angew.Chem.intern. Edn., 1982, 21, 766).

(ii) Reactions

Pyrene forms a radical cation on anodic oxidation (C. Krohnke, V.Enkelmann and G.Wegner, *ibid.*, 1980, 19, 912). On the other hand, reaction with alkali metals produces a dianion **(35)** and a tetra-anion **(36)**, each of which can be visualised as having a delocalised peripheral π-system perturbed by an ethylene bridge (M.Rabinovitz and A.Minsky, Pure and appl.Chem., 1982, 54, 1005). Proton nmr studies appear to confirm that **(35)** is paratropic (16π) and **(36)** diatropic (18π).

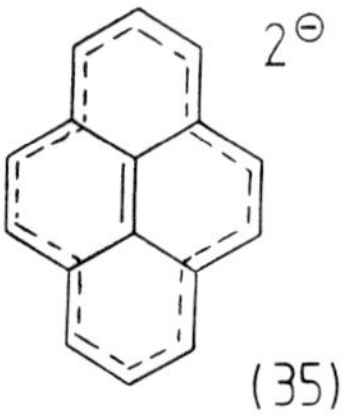

$2^{\ominus}$ (35) $4^{\ominus}$ (36)

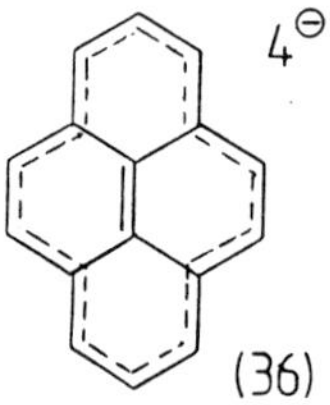

The existence and properties of these ions point to the importance of the peripheral model for π-delocalisation in pyrene (J.R.Platt, J.chem.Phys., 1956, 22, 1448; M.J.S.Dewar, J.Amer.chem.Soc., 1952, 74, 3345; R.Breslow, R.Grubbs and S.I.Murahashi, *ibid.*, 1970, 92, 4139) as opposed to the component model (I.Gutman, M.Milun and N. Trinajstic, *ibid.*, 1977, 99, 1692; M.Randic, *ibid.*, 1977, 99, 445).

Alkylation of the dianion **(35)** in liquid ammonia gives the isomers **(37)** and **(38)** in 26% and 48% yield respectively (C.Tintel *et al.*, Synth.Comm., 1985, 15, 91).

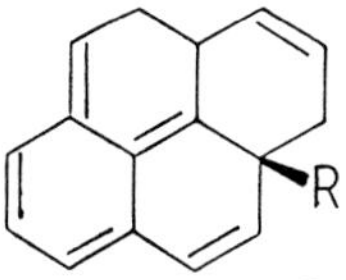

(R = $PhCH_2$)

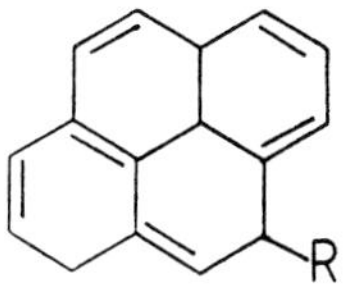

(37) (38)

Pyrene, when irradiated in the presence of an amine and carbon dioxide in a dipolar aprotic solvent, undergoes reductive carboxylation (S.Tazuko, S.Kazama and N.Kitamura, J.org.Chem., 1986, 51, 4548). The product **(39)** is obtained in trace quantities only.

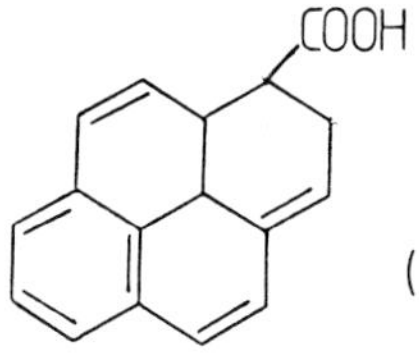

COOH

(39)

The essential principles of electrophilic substitution in pyrene are covered in the 2nd Edition IIIH, p.268; few reactions of significance have since emerged. Diacylation has been reinvestigated (R.G.Harvey, J.Pataki and H.Lee, Org.prep.Proced.Intern., 1984, 16, 144) and di-iodination with $I_2/HIO_3/H_2SO_4$ reported (V.Chaikovskii and A.N.Novikov, J.org.Chem.USSR, 1984, 20, 1350).

The employment of regiospecific hydrogenation in order to prepare isomers which cannot be obtained by reaction of pyrene itself is exemplified by a synthesis of 2,7-dibromopyrene **(40)** (H.Lee and R.G.Harvey, J.org.Chem., 1986, 51, 2847).

Br$_2$, FeCl$_3$ in H$_2$O → ... Br$_2$, CS$_2$ → (40)

The use of bromine as an aromatising agent for the last step in the above synthesis is worthy of note; conventional reagents such as DDQ are ineffective in securing a pure product.

Nucleophilic substitution of pyrene can be achieved by irradiation in the presence of a nucleophile (C.Tintel, J.Cornelisse and F.J.Rietmeyer, Recl:J.R. Neth.chem.Soc., 1983, 102, 224). Thus may be obtained 1-cyano and 1-aminopyrene. Oxygen-containing nucleophiles, which fail to react with pyrene, do react with 1-fluoropyrene to yield, for example, 1-hydroxy- and 1-methoxypyrene. Replacement of fluorine is further exemplified by the reactions of decafluoropyrene, which reacts with methoxide ion at position 1, then 3,6 and 8 (J.Burdon, I.W.Parsons and H.S.Gill, J.chem. Soc., Perkin I, 1979, 1351). The structures of the products of these reactions have been established by nmr spectroscopy and by chemical means, and their formation rationalised on the basis of a recent modification of the I_{π} repulsion theory.

4-Acetoxypyrene has been used as a mechanistic probe to study the reaction of CF_3OF with aromatic compounds. Isolation of a small quantity of the intermediate (41) indicates that monofluorination with this reagent occurs by an addition-elimination mechanism. (T.B.Patrick, G.L. Cantrell and C.Y.Chang, J.Amer.chem.Soc., 1979, 101, 7434).

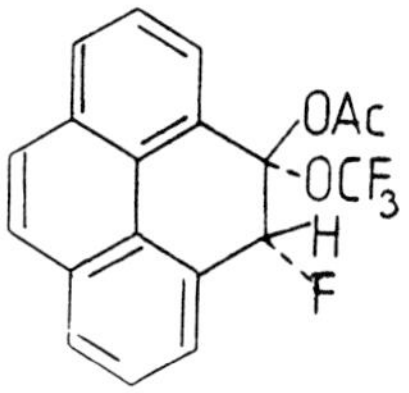
OAc
OCF3
H
F

(41)

(iii) Derivatives

With the ultimate goal of synthesising the cross-conjugated 2,7-dihydropyrene **(42)**, J. Ackerman *et al.* (Angew.Chem. intern.Edn., 1982, 21, 618) have prepared the blocked and sterically-shielded tetramethyl derivative **(43)** in five steps from naphthalene-1,4,5,8-tetracarboxylic acid dianhydride. Although stable in the dark, **(43)** undergoes valence tautomerisation in diffuse daylight to give the isomers **(44)** and **(45)**.

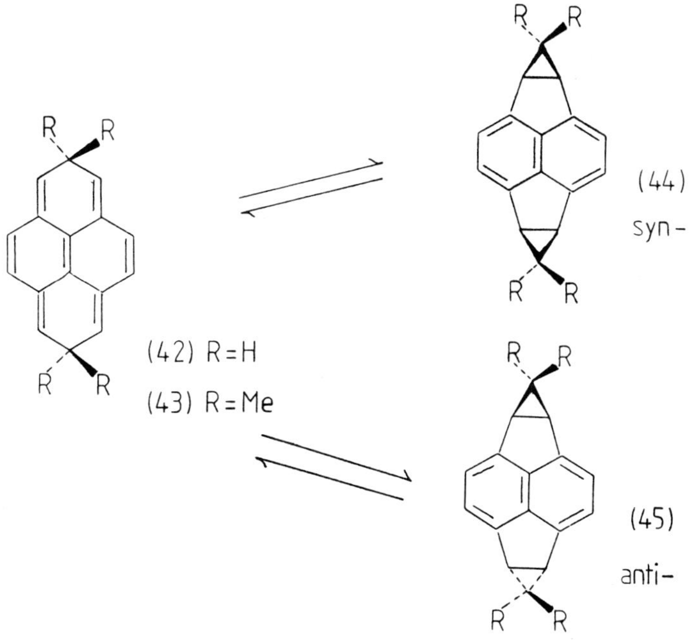
R
(42) R=H
(43) R=Me
(44)
syn-
(45)
anti-

The synthesis of the related pyrene derivative **(46)**, a proposed electron-acceptor for molecular metals analogous to TCNQ (2nd Edn., Vol.IIIB, p.144), has been attempted by two groups of workers (N.Acton *et al*., J.org.Chem., 1982, 47, 1011; M.Maxfield, A.N.Block and D.O.Cowan, *ibid*., 1985, 50, 1789), and success claimed by the latter group.

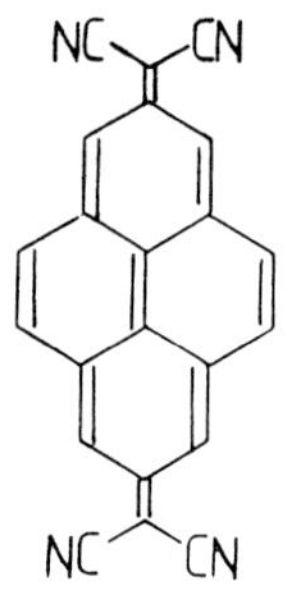

(46)

Several cyclophanes containing the pyrene moiety have been synthesised in order to correlate the structure of the pyrene excimer to the excimer fluorescence (T.Kawashima *et al*., Tetrahedron Letters, 1978, 51, 5115; H.A.Staab and R.G.H.Kirrstetter, Ann., 1979, 886; Staab *et al*., Ber. 1984, 117, 246). An example, [2.2] (2,7) pyrenophane, which has two pyrene units fixed in an ecliptic face-to-face orientation with parallel pyrene axes, is illustrated below.

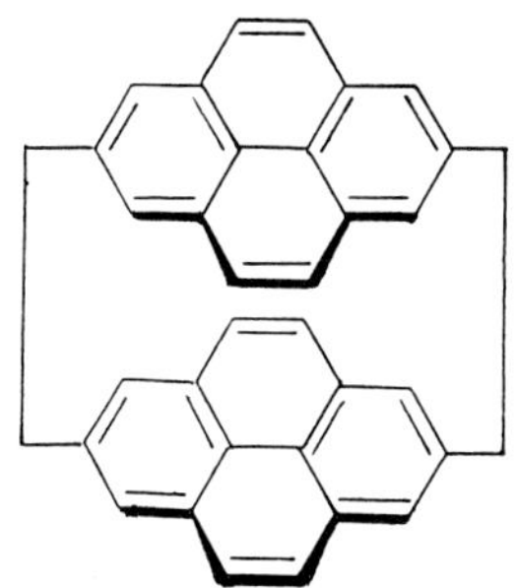

Other isomers which have been prepared include [2,2][1,6]-, [2.2](1,6)(2,7)-, [2.2](1,8)-, [3.3](2,7)- and [4.4](2,7)-pyrenophanes.

(g) 10b,10c-Dimethyl-10b,10c-dihydropyrene

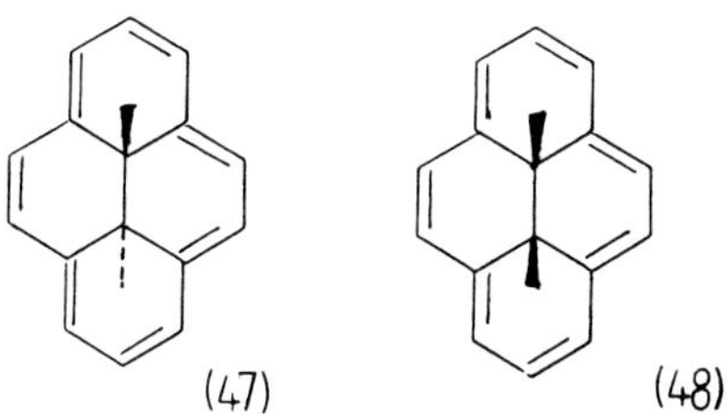

(i) Synthesis

An attempt to increase the *syn/anti* ratio of the dithiacyclophane intermediates (which determines the ratio of (48) to (47) in the final product), by incorporating a nitro group which could be subsequently removed, has met with limited success (D.Kamp and V.Boekelheide, J.org.Chem., 1978, 43, 3475).

An attractive new route to the synthons for the established synthesis (R.H.Mitchell and V.Boekelheide, J.Amer.chem.Soc., 1974, 96, 1574) has been devised (Mitchell *et al*., Tetrahedron, 1986, 42, 1741).

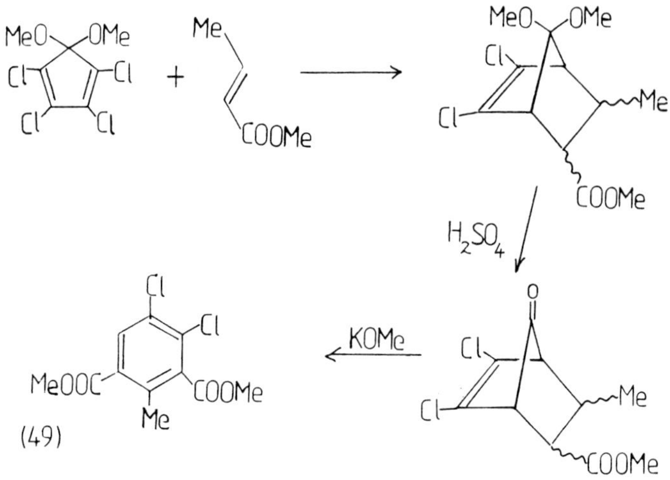

MeO OMe Cl Cl Cl Cl + Me COOMe → MeO OMe Cl Cl Me COOMe

H_2SO_4

O Cl Cl Me COOMe

KOMe

Cl Cl MeOOC COOMe Me

(49)

Conventional transformations on **(49)** lead to the required synthons **(50)** and **(51)**.

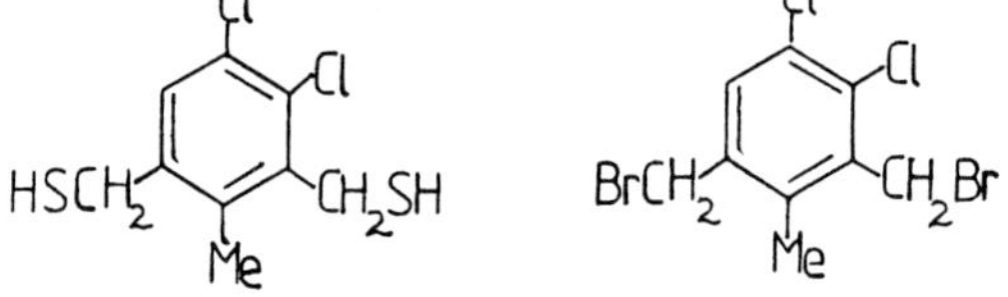

Cl Cl $HSCH_2$ CH_2SH Me

(50)

Cl Cl $BrCH_2$ CH_2Br Me

(51)

An important consequence of the use of these precursors is the dramatic increase in the *cis/trans* ratio of the final product, rendering the *cis*-series more accessible than hitherto.

(ii) Reactions

The reactions of the *cis*-isomer (48) have been found to differ from those of the better-known *trans*-isomer (47). For example, (48) on acetylation yields a mixture of 1- and the 2- acetyl derivative in the ratio 2 : 1, whilst (47) gives the 2-isomer only. The *cis* series reacts more readily with oxygen than the *trans*-series, affording non-aromatic diepoxides (Kamp and Boeckelheide, *loc.cit.*).

(iii) Other 10b,10c-dihydropyrenes

The *trans*-10b-*n*-butyl-10c-methyl derivative has been synthesised using a modification of the established route (T.D.Harris, B.Neuschwander and V.Boekelheide, J.org.Chem., 1978, 43, 727). Also, representing the first example of a dihydropyrene possessing functionality in the cavity of the π-electron cloud, the 10b-(but-1-en-4-yl)-10c-methyl compound has been prepared (Y.Mao and Boekelheide, *ibid*., 1980, 45, 2746). Significantly, the pmr signal of the CH_3 group in the latter is in the usual position, indicating that there is little, if any, interaction between the peripheral π-cloud and the π-electrons of the vinyl group. Likewise, neither an *n*-butyl group nor an ether substituent, $-CH_2.CH_2.OCH_3$, has any effect on the ring current.

A phenyl substituent, being more rigid than an alkyl group and amenable to functionalisation, is an attractive proposition for the study of electronic interactions. *trans*-10b-Methyl-10c-phenyl-10b,10c-dihydropyrene, green crystals, m.p. 159-160°, provides such an opportunity. (R.H. Mitchell and W.Anker, Tetrahedron Letters,1981, 22, 5139). Bond geometry determines that the phenyl ring lies at right angles to the main ring system, so that one aromatic π cloud lies within and perpendicular to the other. Interestingly, whilst the CH_3 protons resonate at a "normal" position (δ -4.45) in the pmr spectrum, the *ortho*-protons of the phenyl group appear at δ 2.55, the most shielded aryl protons reported to date. (*m*-, δ 5.6; *p*-, δ 6.0). It is not known whether the phenyl group is rotating or fixed symmetrically.

Two molecules of the *trans*-dimethyl compound (47) have been coupled *via* the 2-bromoderivative, using *bis*(triphenylphosphino) nickel(II) chloride, to give a macroscopic biphenyl containing the dihydropyrene nucleus (Mitchell *et al*., J.Amer.chem.Soc., 1984, 106, 7776).

(h) *Triphenylene and its derivatives*

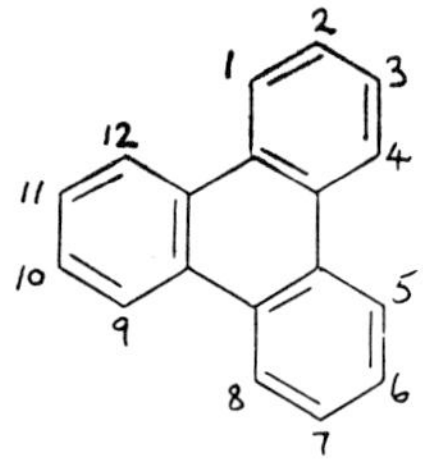

1 2 3 4 5 6 7 8 9 10 11 12

Interest in the triphenylene system, declining over the last few years, has been rekindled with the recognition of disc-shaped molecules as potential mesogens. This has opened up a new field of liquid crystal research. Many thermotropic and lyotropic discoidal mesogens are hexasubstituted triphenylenes, for example **(53)**.

It is formed from the terphenyl **(52)** by oxidative photocyclisation, in a transformation which is conveniently regiospecific.

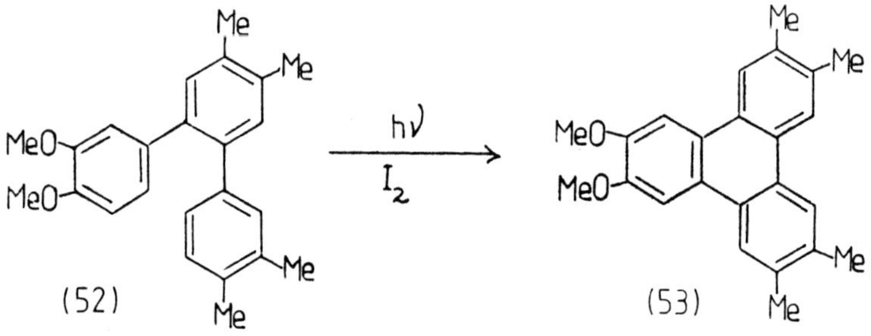

MeO MeO Me Me Me Me (52) $\xrightarrow[I_2]{h\nu}$ MeO MeO Me Me Me Me (53)

The precusor (52) is synthesised by a sequence of standard Ullmann reactions (R.J.Busby and C.Hardy, J.chem.Soc., Perkin I, 1986, 721).

Two new syntheses of the triphenylene system are worthy of mention:

Aryne addition — bridge extrusion provides alkyl-substituted triphenylenes (G.W.Gribble, R.B.Perni and K.D.Onan, J.org. Chem., 1985, 50, 2934).

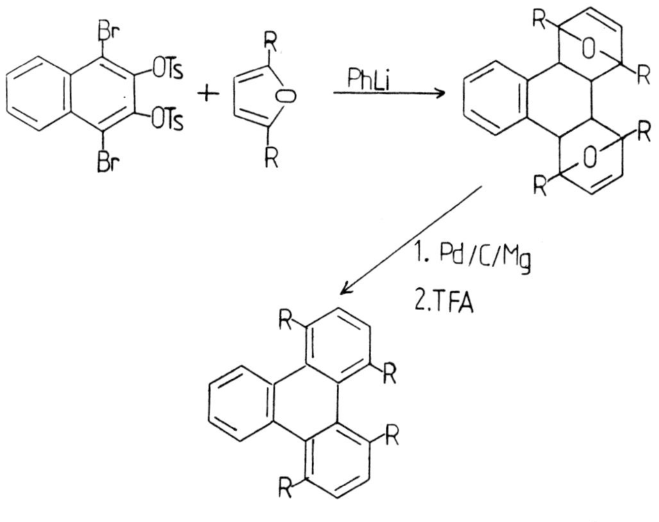

Br, OTs, OTs, Br + R, O, R —PhLi→ R, O, R, R, O, R —1. Pd/C/Mg 2.TFA→ R, R, R, R

Cyclodehydration of a divinylalcohol, obtained from the corresponding quinone, is a general method which is equally applicable to the synthesis of several other polycyclic hydrocarbons (K.B.Sukumaran and R.G.Harvey, J.org. Chem., 1981, 46, 2740).

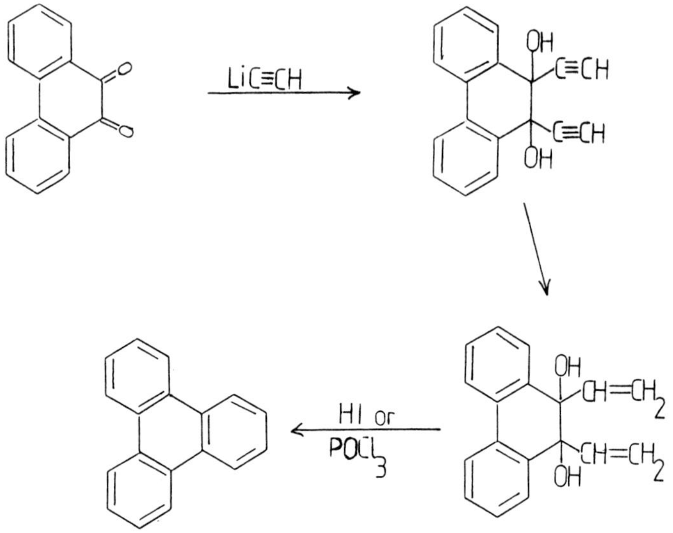

LiC≡CH

HI or $POCl_3$

3 Hydrocarbons containing five benzene rings

Interest in the polycyclic aromatic hydrocarbons in this category has centred mainly on those with suspected carcinogenic properties, almost to the exclusion of new research on others such as pentacene and picene. Omission of a particular hydrocarbon from the present review may be taken to indicate that the treatment in the 2nd Edition is an adequate, if not exhaustive, summary of the current position.

(a) Pentaphene

A new synthesis of this hydrocarbon employs a novel variation of the aryne cycloaddition methodology described elsewhere (R.Camezind and B.Rickborn, J.org.Chem., 1986, 51, 1914). A silyl-protected isobenzofuran is added to 1,2-anthracyne, formed *in situ* from 1-chloroanthracene, and the adduct **(54)** reduced in a single step to pentaphene.

Cl
LTMP*
$SiMe_3$
+
$SiMe_3$
Me_3Si
O
$SiMe_3$
(54)
Pentaphene

* Lithium tetramethylpiperidide

(b) Dibenzanthracenes

(i) Dibenz[a,c]anthracene

A *bis*-Wittig reaction provides a simple route to this compound (A. Minsky and M.Rabinowitz, Synthesis, 1983, 497).

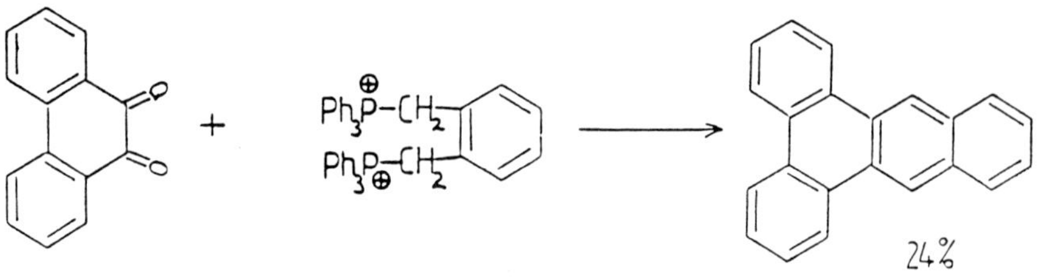

It is also obtained by appropriate modification of the divinyldiol synthesis of triphenylene (p106), a result which illustrates that dibenz[a,c]anthracene is a benzo-fused triphenylene (Sukumaran and Harvey, *loc.cit* .).

A more conventional synthesis involving cyclodehydration and reduction of a keto-acid is given by Harvey *et al*., (J.org. Chem., 1978, 43, 3423), who have also prepared the dihydrodiol epoxide (*ibid*., 1980, 45, 169).

(ii) Dibenz[a,j]anthracene

Syntheses include modifications of established methods (P.Studt, Ann., 1978, 2105; K.L.Platt and F.Oesch, J.org.Chem., 1981, 46, 2601), cycloaddition of styrene to phenanthra-1,4-quinones (G.M.Muschik, T.P.Kelly and W.B. Manning, *ibid*., 1982, 47, 4709), and application of Harvey's *o*-lithioamide method to 2-naphthaldehyde (*cf*. benz[a]anthracene, p.88).

F.Diederich, K.Schneider and H.A.Staab (Ber., 1984, 117, 1255), in a search for a better route to dibenz[a,j]anthracene in connection with a kekulene synthesis (see p.129), observed that photo-cyclisation of 1,3-distyrylbenzene and of 2-styrylnaphthalene yield benzo[c]chrysene rather than the required hydrocarbon. These results are in accord with the rule that, in the case of competing cyclisations, the preferentially-formed product is the one for which the sum of the free valence indices in the excited state of the two atoms involved is the highest. For the dihydro-compound

(55), the two sums are roughly the same, and in the event the product obtained from photolysis of **(55)** is a mixture of **(56)** and **(57)**, from which benzo[c]chrysene **(58)** and dibenz[a,j]anthracene **(59)** (10:6) are obtained.

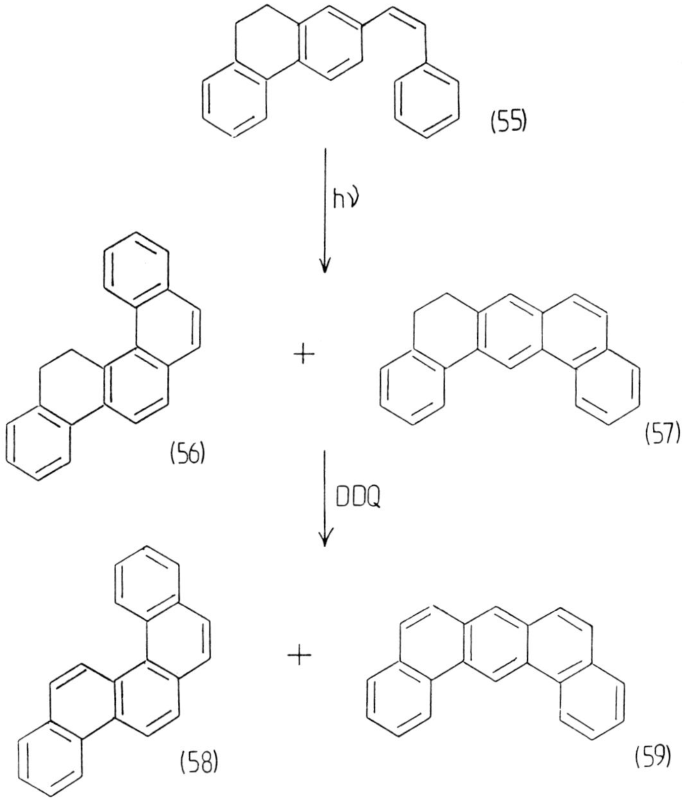

(iii) Dibenz[a,h]anthracene

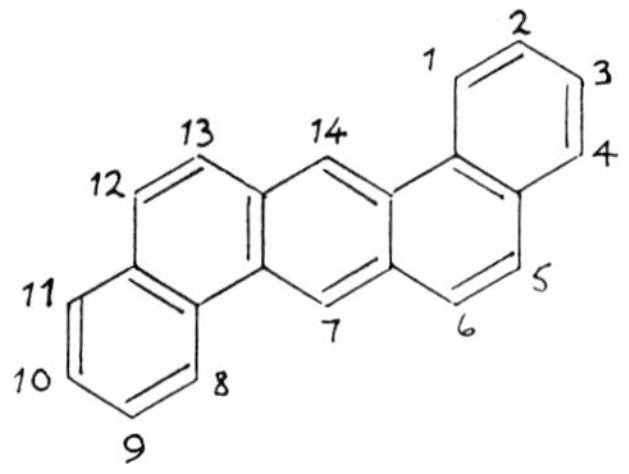

This hydrocarbon, which is a powerful carcinogen, is readily available by established syntheses or their modifications (Platt and Oesch, *loc.cit.*; Muschik, Kelly and Manning, *loc.cit.*). R.G.Harvey's *o*-lithioamide route (p.88) is also applicable, using 1-naphthaldehyde as the precursor. Preparations of the oxygenated metabolites of dibenz[a,h]anthracene are detailed by J.M.Karle *et al.*, Tetrahedron Letters, 1977, 46, 4021; H.Lee and R.G.Harvey, J.org.Chem., 1980, 45, 588; and F.Oesch *et al.*, *ibid.*, 1982, 47, 568.

(c) Benzochrysenes

Benzo[g]chrysene **(61)** has been synthesised from the quinone **(60)** by the divinylalcohol cyclodehydration route (pp.106, 108).

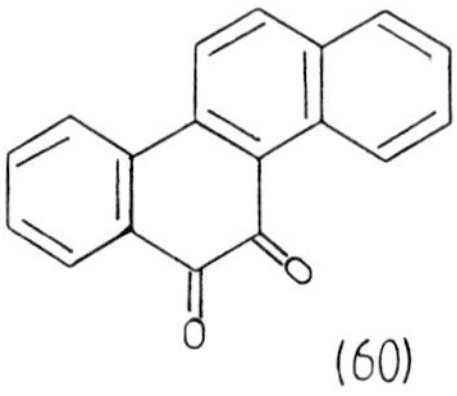

(60) 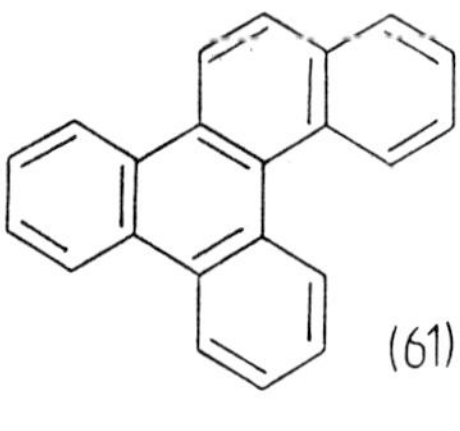 (61)

Benzo[c]chrysene has been referred to above (p.108).

(d) Dibenzophenanthrene

Dibenzo[c,g]phenanthrene **(64)**, now commonly referred to as [5]helicene, attracts interest as the lowest member of the helicene series of non-planar aromatic hydrocarbons. As an alternative to established syntheses, it has been formed *via* its 3,4-dihydroderivative **(63)** by irradiation of the dinaphthylethene **(62)** (R.Lapouyade *et al.*, J.org.Chem., 1982, 47, 1361).

(62) $\xrightarrow[RNH_2]{h\nu}$ (63) $\longrightarrow$ (64)

1-Fluoro[5]helicene has also been synthesised (*idem*., Angew.Chem.intern.Edn., 1982, 21, 766).

The crystal structures of [5]helicene, which exists in two racemic modifications, have been determined (R.Kuroda, J.chem.Soc., Perkin II, 1982, 789) as have those of the racemate and S-(+)-enantiomer of the dihydro derivative **(63)** (R.Kuroda and S.F.Mason, J.chem.Soc., Perkin II, 1981, 870).

(e) Benzopyrenes

The chemistry and cancer-causing properties of the benzopyrenes have been reviewed ("Benzopyrenes", M.R.Osborne and N.T.Crosby, Cambridge University Press, 1987).

(i) Benzo[a]pyrene

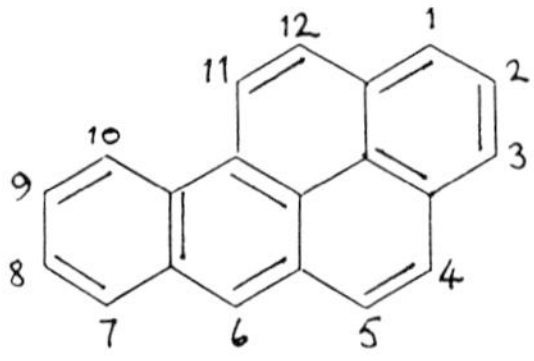

Benzo[a]pyrene

The most powerful carcinogen in coal-tar, benzo[a]pyrene has received more attention than any other polycyclic aromatic hydrocarbon, with the possible exception of benz[a]anthracene.

Two new syntheses of this ring system have been described. The ketone **(65)**, prepared from the corresponding carboxylic acid, is cyclised smoothly by methanesulphonic acid to give 11-methylbenzo[a]pyrene **(66)**.

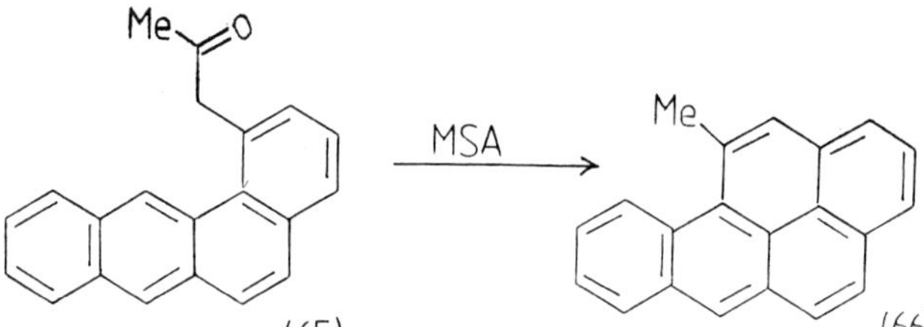

MSA

(65) (66)

The 5-isomer is similarly prepared (R.S.Bodine and G.H.Daub, J.org.Chem., 1979, 44, 4461).

The *o*-lithioamide route (p.88) applied to the ketone **(67)** gives benzo[a]pyrene itself *via* the lactone **(68)** (R.G. Harvey, C.Cortez and S.A.Jacobs, *ibid.*, 1982, 47, 2120).

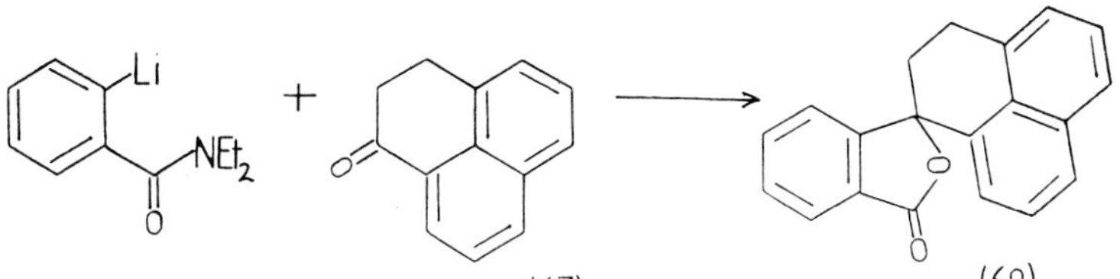

(67) (68)

With appropriate modifications, this route can be used to obtain alkyl derivatives such as 9-*t*-butylbenzo[a]pyrene (J.Pataki, K.Konieczny and R.G.Harvey, *ibid.*, 1982, 47, 1133).

Many other substituted benzo[a]pyrenes have been prepared, some by total synthesis and others by modification of the parent hydrocarbon or its derivatives.
Hydroxy- (P.P.Fu and Harvey, Org.prep.Proced.Intern., 1981, 13, 152; 1982, 14, 414; J.org.Chem., 1983, 48, 1534).
Fluoro- (H.Paulsen and W.Luttke, Ber., 1979, 112, 2907; M.S.Newman and R.Kannan, J.org.Chem., 1979, 44, 3388 ; Newman, V.K.Khanna and K.Kanakarajan, J.Amer.chem.Soc., 1979, 101, 6788).
Chloromethyl- (L.M.Deck and G.H.Daub, J.org.Chem., 1983, 48, 3577; A.A.Leon, G.H.Daub and D.L.Van der Jagt, ibid., 1985, 50, 553).
Nitro- (P.P.Fu, J.med.chem., 1984, 27, 1156); E.Johansen, L.K.Sydnes and T.Greibrokk, Acta Chem.Scand., 1984, 38B, 309).

The foregoing compounds have been prepared with a view to obtaining a better understanding of the biological activity of benzo[a]pyrene and for the same reason various oxygenated metabolites have been synthesised (D.R.Boyd *et al*., Tetrahedron Letters, 1978, 28, 2487; H.Yagi and D.M.Jerina, J.Amer.chem.Soc., 1982, 104, 4026; F.Oesch *et al*., Angew.Chem.intern.Edn., 1985, 24, 699; P.P.Fu, C-C.Lai and S.K.Yang, J.org.Chem., 1981, 46, 220).

(ii) Benzo[e]pyrene

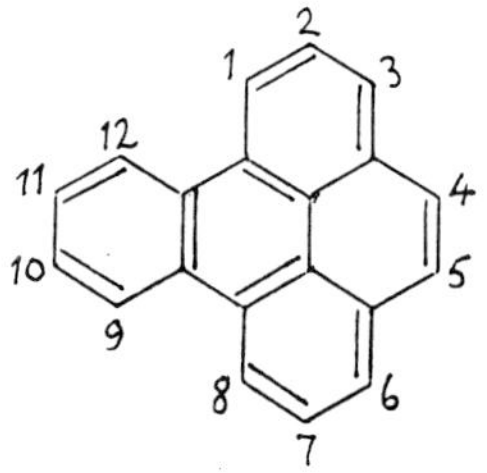

The Jutz synthesis of benzopyrenes from benzanthrene, outlined in the 2nd Edn. IIIH, p.237, was reported as giving a mixture of benzo[a] and benzo[e]pyrenes. In a reappraisal of this reaction, H.Lee and R.G.Harvey (Tetrahedron Letters, 1981, 22, 995) have found it to be regiospecific, affording only benzo[e]pyrene.
Interaction of the pyrene dianion and 1,4-diodobutane yields *cis*-hexahydrobenz[e]pyrene, from which the fully aromatic compound is obtained in 46% overall yield by treatment with DDQ (C.Tintel *et al.*, Chem.Abs., 1983, 99, 53316).
When pyrene-4,5-quinone is used as the substrate in the divinyldiol cyclodehydration procedure of Harvey and Sukumaran (p.106), benzo[e]pyrene is the resulting product.

3-Hydroxybenzo[e]pyrene is readily available *via* direct bromination, but since little is known about the patterns of electrophilic substitution in this hydrocarbon, all the other hydroxy-derivatives have had to be synthesised by individual routes (H.Lee, N.Shyamasundar and R.G.Harvey, J.org.Chem., 1981, 46, 2889), and the same constraints apply to the synthesis of the monomethylbenzo[e]pyrenes (*idem.*, Tetrahedron, 1981, 37, 2563). The K-region and non-K-region *trans*-dihydrodiols of benzo[e]pyrene have been synthesised (R.E.Lehr *et al.*, J.org.Chem., 1978, 43, 3462).

(iii) Dihydropyrene derivatives

trans-12b,12c-Dimethyl-12b,12c-dihydrobenzo[e]pyrene **(69)**, a benzologue of the well-known pyrene derivative **(47)** (p.101) has been synthesised by a simple modification of the established route.

CH_2Br Me Me CH_2Br $\xrightarrow{Na_2S}$ S

1 Wittig rearrangement

2 Hofmann elimination

(69)

The position of the CH_3-resonance in the proton nmr spectrum (δ -1.85) suggests that **(69)** sustains about 55% of the ring-current of the non-benzannelated hydrocarbon **(47)**. This is much greater than that for benz[14]annulenes, and is probably a reflection of the rigidity of the dihydropyrene nucleus. (R.H.Mitchell and J.S-H.Yan, Canad.J.Chem., 1977, 55, 3347; Mitchell, Yan and T.W. Dingle, J.Amer.chem.Soc., 1982, 104, 2551)

Electrophilic substitution of **(69)** (nitration, acetylation) occurs in the 2-position, as in the parent system **(47)**, proving that annelation has no effect on the position of substitution (*idem.*, Tetrahedron Letters, 1979, 15, 1289).

trans-12b,12c-Dihydrobenzo[a]pyrene and its 12b,12c-dimethyl derivative have also been prepared, and the former found to rapidly dehydrogenate to benzo[a]pyrene (R.H.Mitchell *et al.*, J.Amer.chem.Soc., 1982, 104, 2544).

(iv) Benzo[cd]pyrene, naphthanthrene

Although the isolation of 6-methylenebenzo[cd]pyrene **(70)** has not been accomplished, there is considerable evidence from trapping reactions and the formation of dimerisation products that it is formed as a reactive intermediate (O.Hara, K.Yamamoto and I.Murata, Bull.chem.Soc.Japan, 1980, 53, 2036)

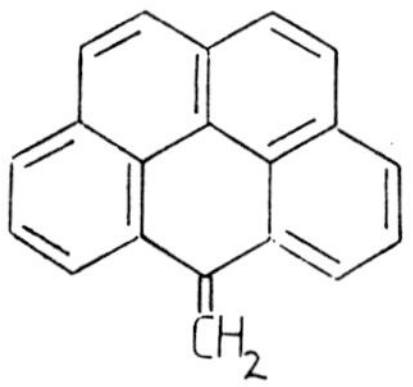

CH_2

(70)

(f) Perylene and its derivatives

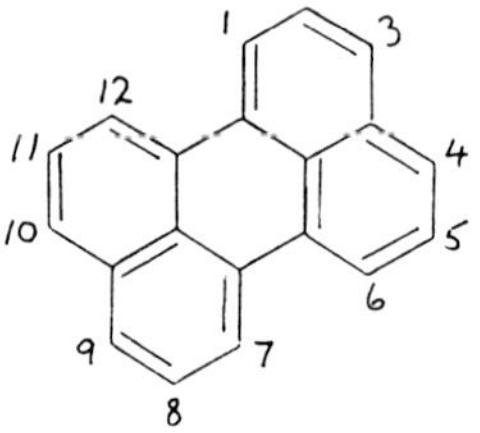

1 2 3 4 5 6 7 8 9 10 11 12

No new syntheses of this hydrocarbon have been reported, and its chemistry appears to have been neglected in the past decade. Nitration of perylene with dilute nitric acid has been reported as giving 3-nitro- and 1-nitro-perylene in the ratio 56:24 (2nd Edn. III H, p.294). A recent study, however, quotes a 93:3 ratio. A large excess of nitric acid affords a mixture of dinitroperylenes, partial separation of which by hplc gives 3,6-dinitro-, 3,7-dinitro, and a mixture of the 3,9- and the 3,10-isomer (A.Nordbotten, L.K.Sydnes and T.Greibrokk, Acta.Chem.Scand., 1984, 38B, 701). Use of the nitrating agent N_2O_4 results in the almost exclusive formation of 3-nitroperylene (F.Radner, *ibid.*, 1983, 37, 65).

Several fungal pigments, closely related to elsinochrome (2nd Edn. IIIH, p.297), have been isolated from moulds. Hypocrellin (71), whose structure has been established by proton and ^{13}C nmr, and confirmed by X-ray crystallography (Chen Wei-shin *et al.*, Ann., 1981, 1880) is found in the fungus *Hypocrella bambusae*.

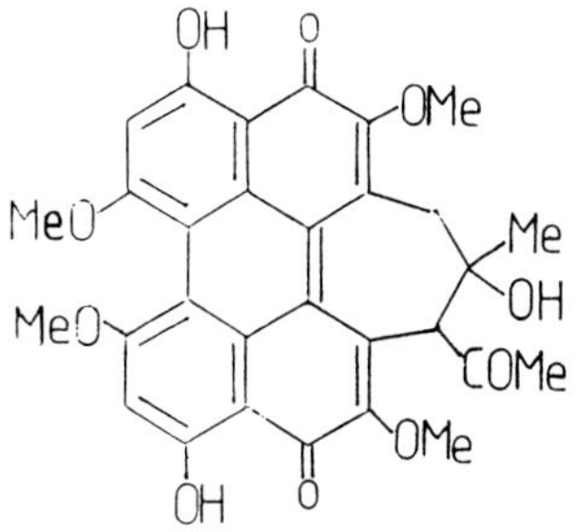

(71)

The novel amphicercosporin (72a), reddish needles m.p.195-6°, protocercosporin (73), orange orthorhombic crystals m.p.260° and neocercosporin (72b), m.p.237°, are obtained together with the known cercosporin (72c), from *Cercospora Kikuchii var.* (S.Matsueda *et al.*, Chem.and Ind., 1982, 58)

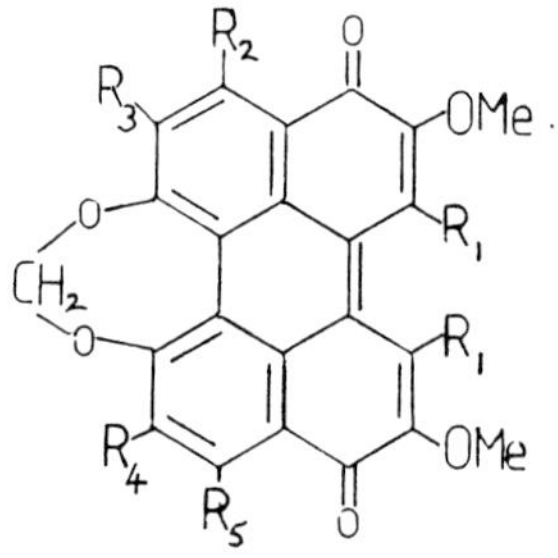

(72)

72a : R_1 = $CH_2CH(OH)CH_3$; R_2 = R_4 = OH; R_3 = R_5 = H
72b : R_1 = $CH_2CH(OH)CH_3$; R_3 = R_4 = OH; R_2 = R_5 = H
72c : R_1 = $CH_2CH(OH)CH_3$; R_2 = R_5 = OH; R_3 = R_4 = H

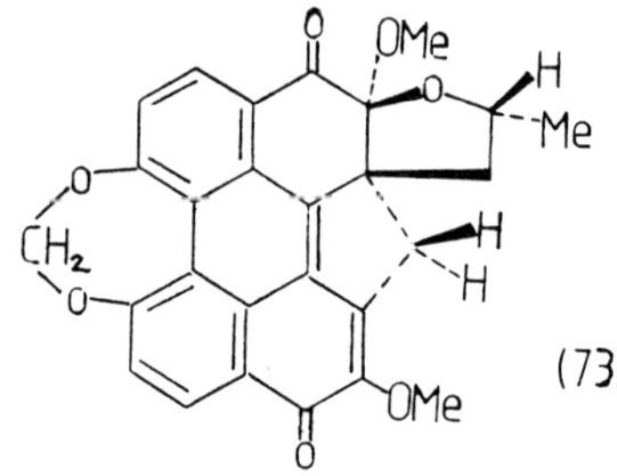

(73)

The structures of these pigments follow from their uv-, ir- and pmr-spectra, and the configuration of (73) has been established by X-ray crystallographic analysis.

*(g) Triptycene**

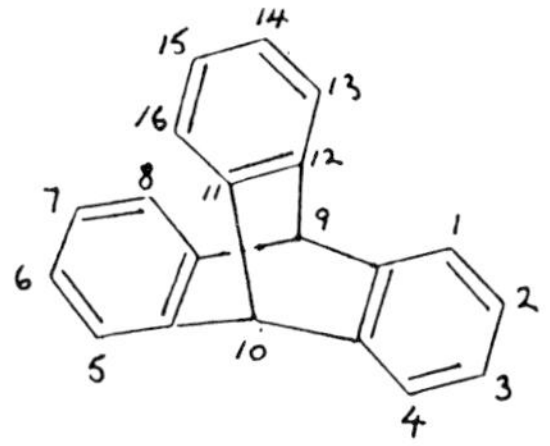

Triptycene

Since its first preparation in 1942, the rigid framework of triptycene has proved attractive for the study of such diverse phenomena as intramolecular charge transfer and restricted rotation about single bonds. (For relevant references, see H.Hart, S.Shamouilian and Y.Takehira, J.org.Chem., 1981, 46, 4427). The same authors have proposed a generalisation of the triptycene concept, suggesting the name "iptycenes", for example heptiptycene (74).

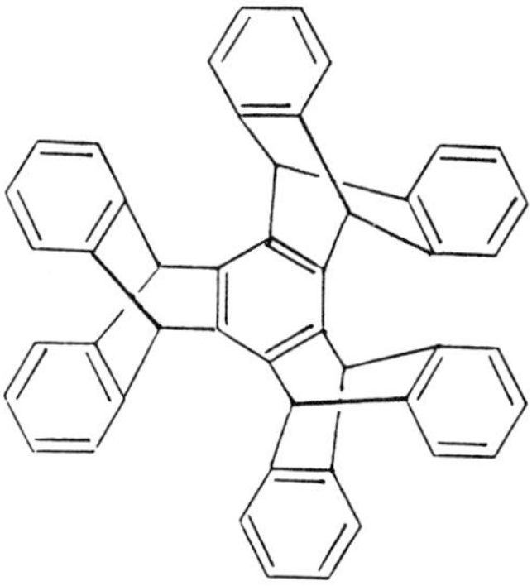

(74)

* Note: Different numbering systems are in use. The bridgehead position (9-) is sometimes numbered as "1-".

Unusual regioselectivity is observed in the alkali metal/ammonia reduction of triptycene. Potassium in THF gives the phenylanthracene derivative **(75)**, but lithium in *t*-butanol produces a mixture of the dihydrotriptycenes **(76)** and **(77)**, the latter unexpectedly predominating (P.W. Rabideau *et al.*, *ibid.*, 1979, 44, 4594).

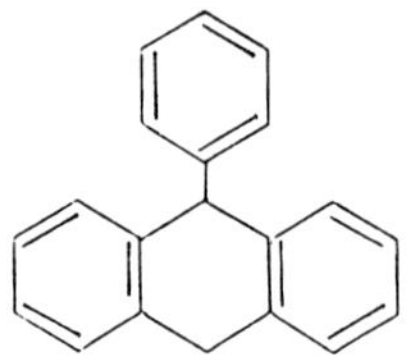

(75) 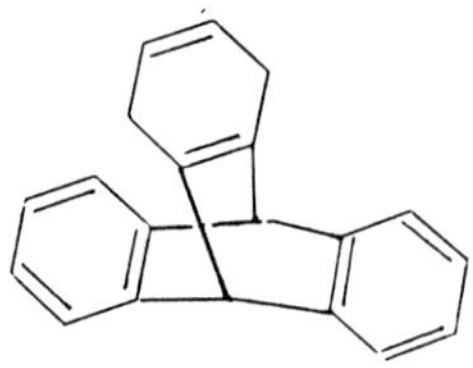 (76) 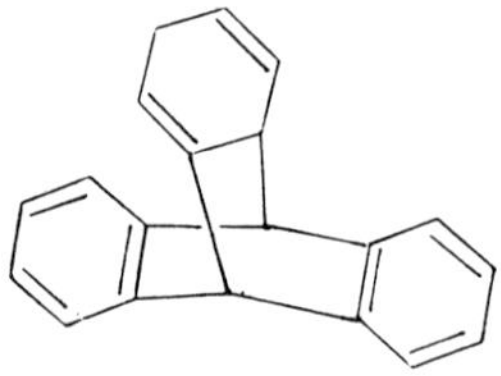 (77)

The unusual steric nature of the triptycene molecule may be responsible for the remarkable stability of its 9-sulphenic acid **(78**, X= -S-OH), m.p.204-205° (N.Nakamura, J.Amer.chem. Soc., 1983, 105, 7172).

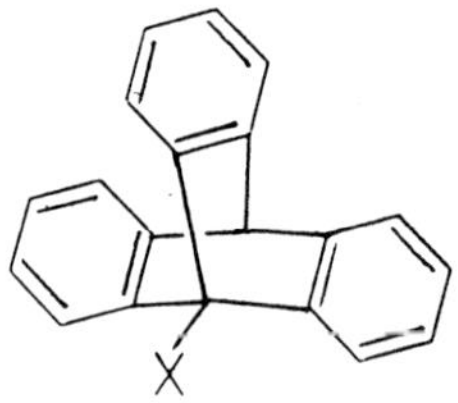

X (78)

The structure of the sulphenic acid is supported by a signal at 65.11 ppm in its ^{13}C-nmr spectrum (a characteristic feature of a triptycene bridgehead carbon bearing a divalent sulphur functionality) and by other spectral data. This result points to an effective method for the stabilisation of other labile functional groups.

Photolysis of triptycenes possessing a leaving group at a bridgehead position **(78**, X= -OCOPh, -SCH, -OPh) results in the formation of benz[a]indeno[1,2,3-cd]azulene **(80)**, possibly *via* the intermediate **(79)** (Y.Kawada, H.Tukada and H.Iwamura, Tetrahedron Letters, 1980, 21, 181).

(78) ⟶ [(79)] ⟶ (80)

One of the above results differs from an earlier report that the phenoxy-derivative **(78**, X=OPh), is photolysed to benz[a]aceanthrylene and that alkoxytriptycenes afford mixtures of 9-phenylfluorenes upon irradiation (H.Iwamura and H Tukada, *ibid.*, 1978, 37, 3451). In a study of rotational barriers in 9-isopropyltriptycene **(78**, X=iPr), it emerges that a substituent in a "peri" position relative to the isopropyl group has a considerable influence upon its rotation, as determined by its dynamic-nmr behaviour (G.Yamamoto and M.Ōki, Bull.chem.Soc. Japan, 1983, 56, 2082). (For similar studies on peri-substituted naphthalenes, see 2nd edn., supplement, IIIG, p.179.)

4 Hydrocarbons containing six or more benzene rings

In contrast to the situation in the 4- and 5-ring series, there has been very little systematic development of new syntheses of hydrocarbons containing six or more rings. A short selection is presented below.

An efficient synthesis of dibenzo[b,k]chrysene (81) from the synthons shown, uses the aryne cycloaddition — bridge elimination route (p.85) (C.S.Le Houllier and G.W.Gribble, J.org.Chem., 1983, 48, 1682).

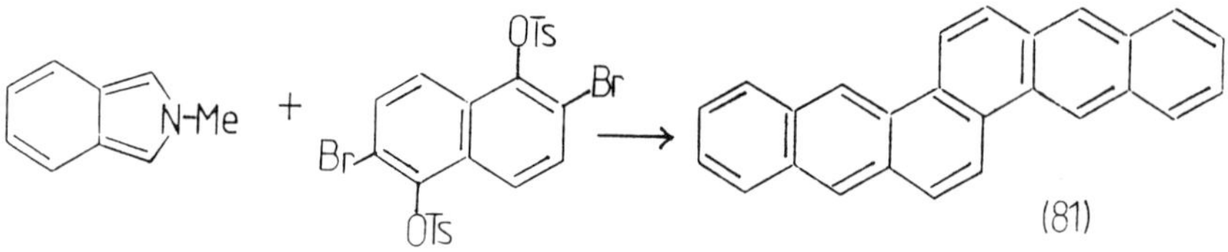

Dehydrocyclisation of *trans,trans*-2,3-distyrylnaphthalene by irradiation produces benzo[s]picene **(82)** (T.S.Skovokhodova, V.I.Luk'yanov and E.B.Merkushev, Chem.Abs., 1980, 92, 6305)

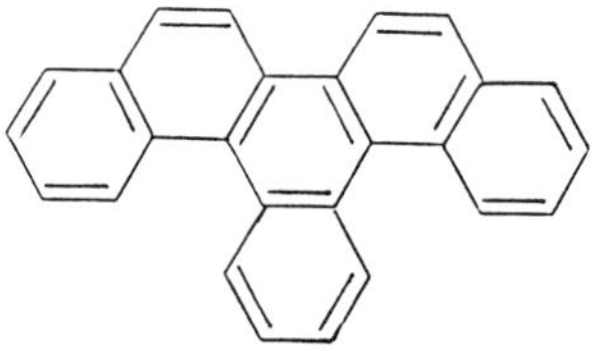

An adaptation of Mitchell and Boekelheide's sulphur-extrusion route to the dihydropyrenes (p.101,115) provides a novel synthesis of zethrene, dibenzo[de,mn]naphthalene **(83)**, from the precursor **(84)**. (W.Kemp, I.T.Storie and C.D. Tulloch, J.chem.Soc., Perkin I, 1980, 2812).

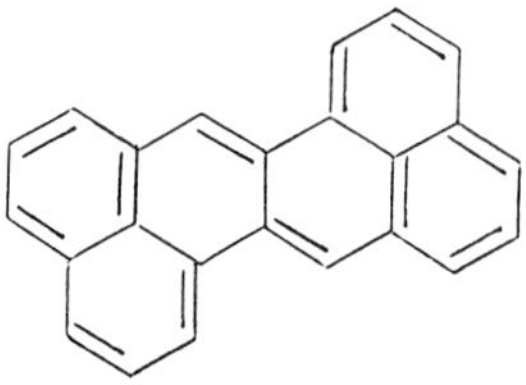

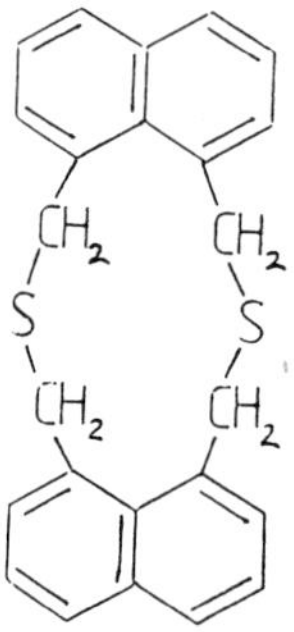

Hexabenzo[a,c,g,i,m,o]triphenylene is formed by flash vacuum pyrolysis of cyclobuta[l]phenanthrene-1,2-dione (J.F.W. McOmie *et al.*, J.chem.Soc. Perkin I, 1982, 19). The product is identical to that obtained previously by J.W.Barton and A.R.Grinham (*ibid.*, 1972, 634), but different to a hydrocarbon obtained by J.G.Carey and I.T.Millar (*ibid.*, 1959, 3144) which has been shown not to have this structure (D.Biermann and W.Schmidt, J.Amer.chem.Soc., 1980, 102, 3163).

A homologue of the phenalenium ion, the cyclohexa[cd]perylenium ion **(85)**, has been synthesised by conventional ring-building and cyclisation from perylene-3-aldehyde (K. Yamamura, H.Miyake and I Murata, J.org.Chem., 1986, 51, 251).

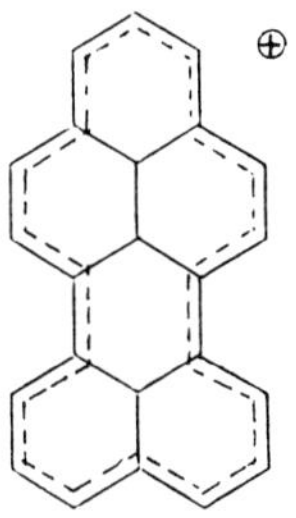

(85) 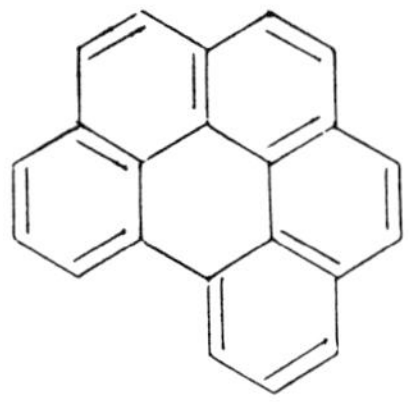 (86)

The tetrafluoroborate of **(85)**, red powder m.p. $>300^{o}$, can be stored without change under atmospheric conditions. Spectral evidence points to a strong diamagnetic ring current associated with the perimeter π-system. Other polycyclic ions which have been prepared include the dibenzo[de,hi]naphthacenyl dication (K.Yamamoto *et al.*, Tetrahedron Letters 1982, 877), and the triangulenyl dianion (O.Hara *et al.*, *ibid.*, 1977, 2435).

The electrophilic substitution of benzo[ghi]perylene **(86)** has been re-examined, the results confirming those reported earlier. Thus, nitration gives the 5-nitro derivative together with some 7-isomer (E.Johansen, L.K.Sydnes and T. Greibrokk, Acta Chem.Scand., 1984, 38B, 309); and acetylation affords the 5-acetyl compound (A.Ivanenko, V.S.Kuznetsov and A.V.El'tsov, Chem.Abs., 1984, 100, 67965).

Helicenes (2nd Edn.IIIH, p.215)

The irradiation of 2-styrylbenzo[c]phenanthrene (**87**) in chiral media might be expected to produce non-racemic [6]helicene (**88**).

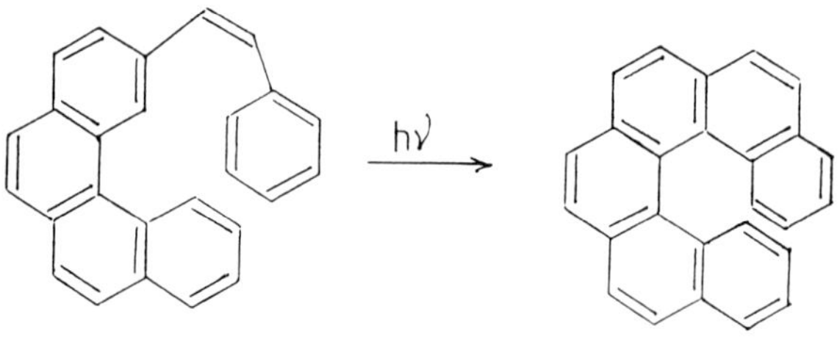

hν

(87)

(88)

This result has been achieved in two ways: Irradiation in eleven different chiral solvents produces non-racemic (**88**) with optical yields of 0.2-2.0% (W.H.Laarhoven and T.J.H.M. Cuppen, J.chem.Soc., Perkin II, 1978, 315), and irradiation in a mechanically-twisted nematic mesophase yields optically active (**88**) whose chirality is governed by the handedness of the twist (M.Nakazaki *et al.*, Chem.Comm., 1979, 1086). The nematic mesophase employed in these experiments was a 1:1 mixture of *p*-cyanophenyl *p*-butylbenzoate and *p*-cyanophenyl *p*-heptylbenzoate, and gave optical yields of 0.4 to 0.9%. The photocyclisation illustrated above has also been investigated to determine its mechanism under anaerobic conditions (Laarhoven, Cuppen and H.H.K.Brinkhof, Tetrahedron, 1982, 38, 3179). The dihydrocompound (**89**) is formed first, but undergoes a 1,5 suprafacial shift to give *trans*-6a,16d-dihydrohexahelicene (**90**).

(87) → (89) → (90) → (88)

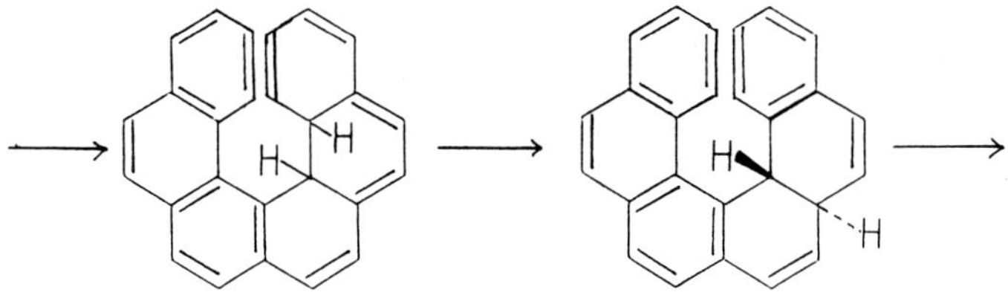

The geometry of [6]helicence offers the possibility of ring-to-ring intramolecular reactions. Examples of these include a carbene insertion from the 1-position (J.Jespers, D.Defay and R.H.Martin, Tetrahedron, 1977, 33, 2141), and the thermally-induced cyclisation of 1,16-dimethyl derivatives (J.H.Borkent, P.H.F.M.Rouwette and W.H.Laarhoven, *ibid.*, 1978, 34, 2569).

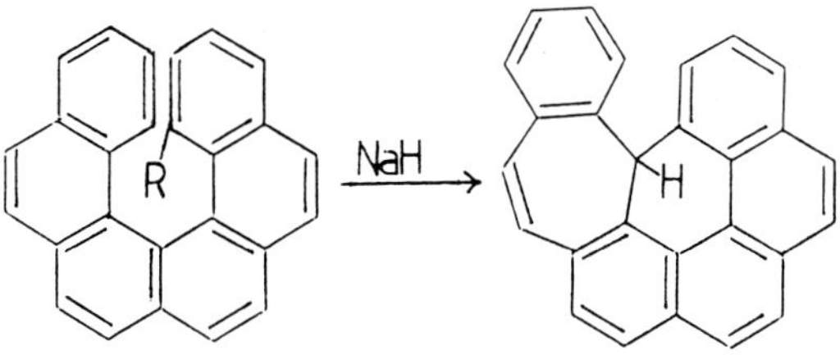

NaH

(R = -CH=NHTs)

Me Me

Δ 180 — 300°

Me

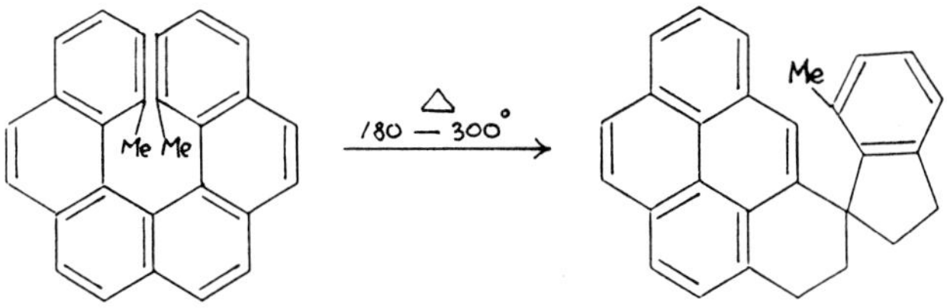

[6]Helicene undergoes addition-elimination reactions at the phenanthrene-type 5,6 and 11,12 bonds. Thus bromine gives adducts which, when heated above 100°, eliminate HBr to afford 5-bromo and 5,12-dibromo derivatives. Analogous products are obtained on nitration and acetylation under mild conditions. (P.M.op den Brouw and W.H.Laarhoven, Rec.Trav.chim., 1978, 97, 265)

Dibenzodihydropyrene derivatives

In a development of their work (p115) on the benzo-annelation of 10b,10c-dimethyl-10b,10c-dihydropyrene, Mitchell and co-workers have synthesised the dibenzo compounds **(91)** and **(92)** (R.Mitchell, R.V.Williams and T.W.Dingle, J.Amer.chem.Soc., 1982, 104, 2560).

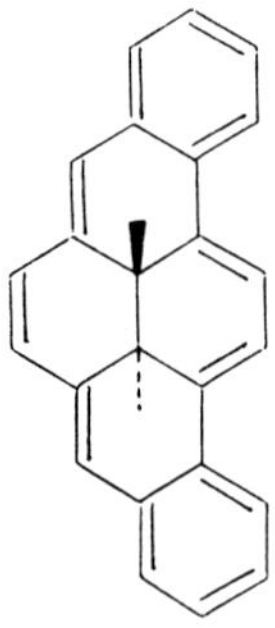

(91) 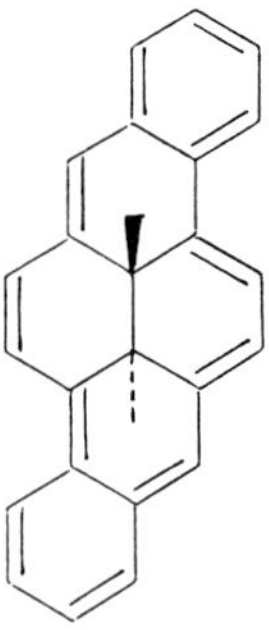 (92a) 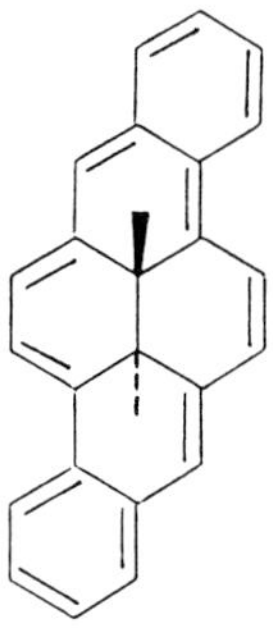 (92b)

A comparison of the pmr chemical shifts of the central methyl groups in the dimethyl dihydropyrene series is most revealing. As expected, progressive benzo-annelation leads to decreased shielding of the methyl groups due to a reduction in the diatropicity of the peripheral ring-system. The dibenzo[a,h] compound **(92)**, however, is a remarkable exception. Its CH_3 groups are as shielded as those of the non-benzannelated compound **(47)**, indicating an undiminished ring-current.These results appear to demonstrate clearly the importance of symmetrical Kekulé structures to diatropicity: Compound **(92)** can be written in two sets of equivalent Kekulé structures **(92a)**, **(92b)**, whereas **(91)** cannot. Hence **(92)** is better considered as a macroscopic annulene than as a dibenzannelated dihydropyrene (Mitchell, R.J.Carruthers and L.Mazuch, *ibid*., 1978, 100, 1007). (For a theoretical treatment of the diatropicity of dihydropyrene and its benzologues in terms of bond order deviations, see R.H.Mitchell *et al*., *ibid*., 1982, 104, 2571). The results confirm that bond localisation caused by benzo-ring annelation is the principal determinant of the strength of the macrocyclic ring current in these compounds. Another example of annelation in this series is provided by the synthesis of the 9,10-phenanthro-annelated derivative (Y-H.Lai, *ibid*., 1985, 107, 6679). In this compound the CH_3 protons resonate at δ -3.32, a result taken to be indicative of the high π-bond order of the 9,10-bond of phenanthrene, which causes much less bond localisation in the 14π macrocyclic ring compared to the benzannelated system **(69)**.

Circulenes

Polynuclear hydrocarbons which consist of a ring of annelated benzene rings have been given the general name "Circulenes" (J.H.Dopper and H.Wynberg, J.org.Chem., 1975, 40, 1957). Two separate categories of circulene may be distinguished: (a) Those having fewer than eight rings, surrounding a central core which can be visualised as an additional fused ring of 5, 6 or 7 carbons, and (b) those in which a cyclic system of more than six benzene rings encloses a cavity lined with hydrogen atoms.

Circulenes in category (a), **(93)**-**(95)**, also referred to in the literature as "corannulenes" and "coronaphenes", have been reviewed in theoretical terms (I.Agranat, Pure and appl.Chem., 1980, 52, 1399).

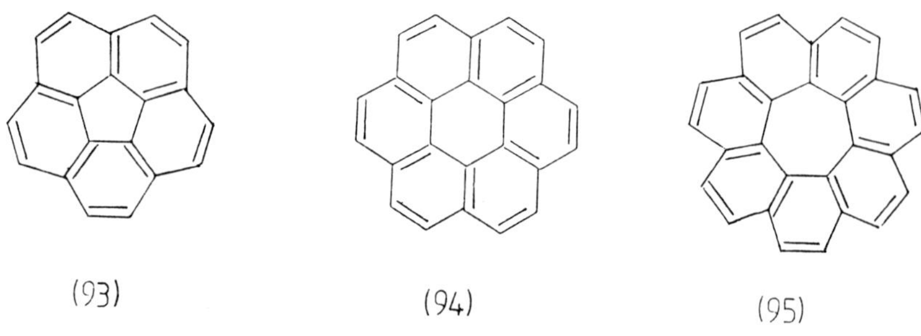

(93) (94) (95)

Whereas [5]circulene (corannulene) **(93)** and [6]circulene (coronene) **(94)** have been known for some time (2nd Edn, IIIH, pp.309-310), the third member of the series, [7]circulene **(95)**, has been reported only recently (K.Yamamoto, T.Harada and M.Nakazaki, J.Amer.chem. Soc., 1983, 105, 7171). The dibromo compound **(96)**, obtained by a multi-stage synthesis, is converted into a dialdehyde and cyclised by reductive coupling with $LiAlH_4/TiCl_3$.

(96)

[7]Circulene, $C_{28}H_{18}$, yellow plates m.p. 295-296°, is a saddle-shaped molecule in which the central 7-membered ring assumes a boat conformation. Its pmr spectrum has a single peak at δ 7.45 and the ^{13}C nmr-spectrum exhibits three peaks at δ 136.0, 132.1 and 127.5. X-ray crystallographic analysis reveals four classes of carbon-carbon bond with average lengths (proceeding outwards from the centre) of 145.7, 143.4, 141.4 and 133.8 pm.

The first, and until recently the only, example of a circulene in category (b) is the [12] circulene **(97)**, which was given the name "kekulene" on the occasion of the centennial of Kekulé's benzene formula (H.A.Staab and F. Diederich, Angew.Chem.intern.Edn., 1978, 17, 372).

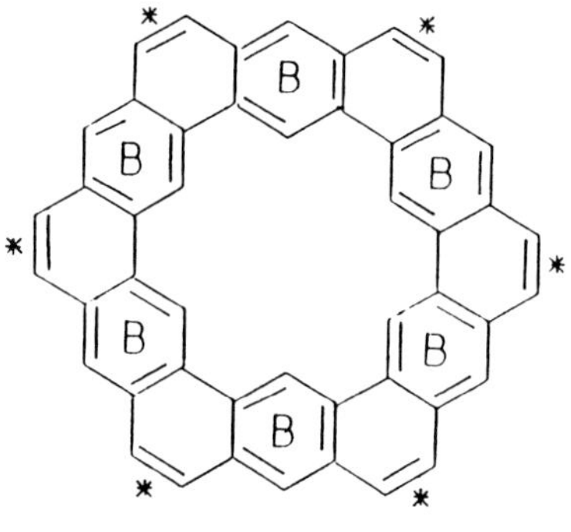

(97a)

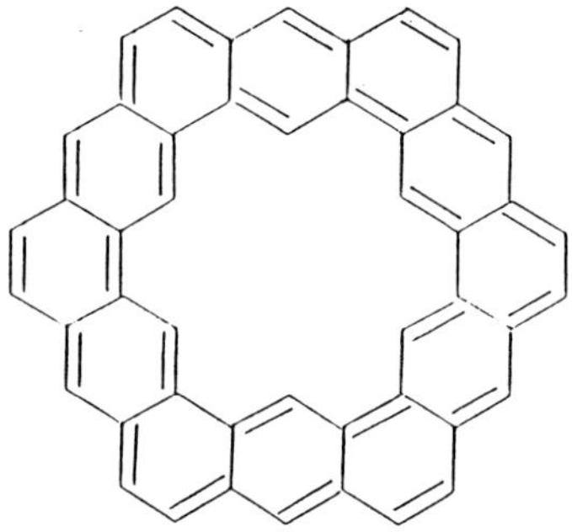

(97b)

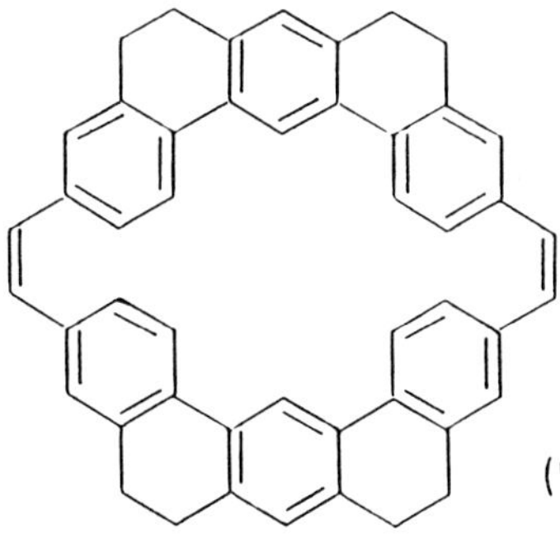

(98)

Kekulene, $C_{48}H_{24}$, greenish yellow microcrystals, m.p. >620°, is an extremely insoluble hydrocarbon. It is synthesised by photolytic cyclisation of **(98)** (prepared from the corresponding dithiacyclophane by established cyclophane procedures), followed by aromatisation of the product with DDQ (Staab and Diederich, *loc.cit.*; *idem.*, Ber., 1983, 116, 3487). The mass spectrum shows a molecular ion at 600, and additional peaks at 300 and 200 representing a dication and trication. Aromatic stretching bands appear in its infrared spectrum at 3000 and 3020 cm^{-1}, but there are few vibrational peaks owing to the high symmetry. The pmr spectrum, obtained with considerable difficulty, exhibited peaks at δ 7.94, 8.37 and 10.45 in the ratio 2:1:1. Clearly there is no upfield signal for the internal protons of a "double annulene" system **(97b)**: The internal protons are probably responsible for the signal at δ 10.45. Thus the "benzenoid" formulation **(97a)** is favoured, a conclusion which is substantiated by X-ray crystallographic analysis. The rings labelled B are evidently benzenoid, with bond lengths 138.8, 139.8 and 142.3 pm falling within the aromatic range, whereas the six bonds marked * are more ethylenic in character with a length of 134.6 pm. (Staab *et al.*, *ibid.*, 1983, 116, 3504)

Nomenclature of circulenes. There are several possible circulenes having twelve fused-rings, so that the name [12]circulene for kekulene is ambiguous. Staab and Diederich (*loc.cit.*) have proposed a logical nomenclature system (*) which is more informative and less clumsy than that used by Chemical Abstracts: Proceeding clockwise through the macrocycle each annelation is related to the previous one by labelling as "a" the bond of fusion between the two rings, then counting the bonds of the individual rings in a clockwise sequence. Thus, for kekulene, a linear annelation is "d", and an angular one "e". Kekulene becomes cyclo[d.e.d.e.d.e.d.e.d.e.d.e.]dodecakisbenzene.

The success of the synthesis of kekulene has prompted many attempts to prepare other circulenes such as the [9]circulene **(99)** (Staab and M.Sauer, Ann., 1984, 742; C.F.Wilcox *et al.*, Tetrahedron Letters, 1978, 22, 1893; F.B.Mallory, C.W.Mallory and S.E.Sen Loeb, *ibid.*, 1985, 26, 3773), and the [10]circulene **(100)** (Staab, Diederich and V.Caplar, Ann., 1983, 2262)

---

* An alternative system has been proposed by I. Agranat, B.A.Hess and L.J. Schaad, Pure and appl.Chem., 1980, 52, 1399)

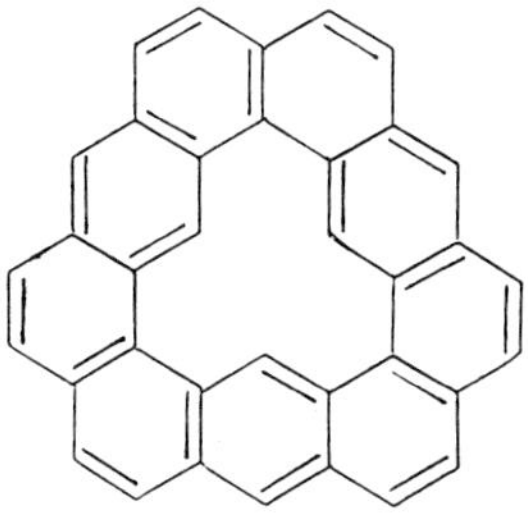

(99)

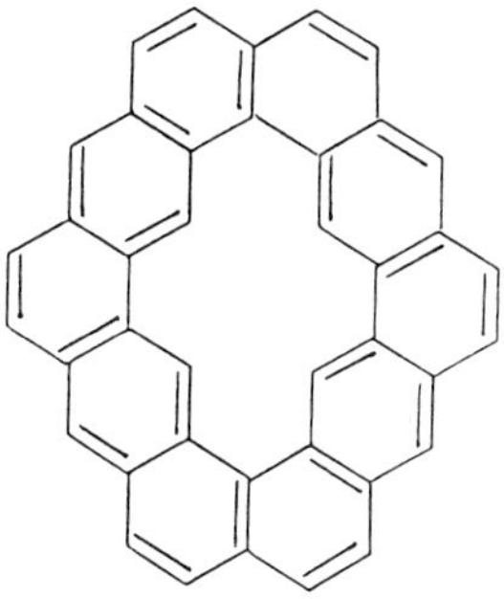

(100)

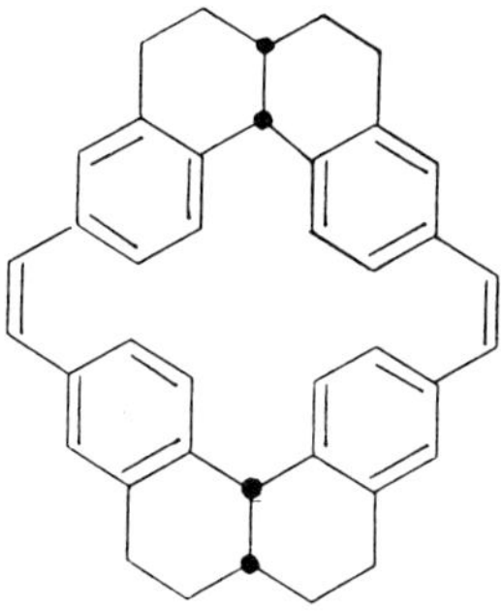

(101)

The failure of these attempts is largely attributable to the resistance to cyclodehydrogenation of the sterically crowded cyclophane precursors in the final step. This impasse has been overcome in the successful synthesis of (100) by D.J.H.Funhoff and Staab (Angew.Chem.intern.Edn., 1986, 25, 742), in which the partially-hydrogenated cyclophane (101) is cyclised by irradiation in *n*-propylamine, and the product aromatised using DDQ.

Cyclo[d.e.d.e.e.d.e.d.e.e]decakisbenzene (100), yellow crystals m.p. >330°dec., has a mass spectrum typical of such polycyclics (M^{+} 500, M^{2+} 250, M^{3+} 166.7, M^{4+} 125) and, interestingly, another peak at M-4 which may be due to loss of the four internal hydrogens with transannular bond formation.

The molecular ion at high resolution appears at 500.1565 (calculated for $C_{40}H_{20}$, 500.1575). Five signals of equal intensity are observed in its pmr spectrum. The low field shift of the inner protons shows that, as with kekulene, no significant diatropism of the macrocyclic system exists.

It is perhaps appropriate that this Chapter should conclude with such apposite illustrations of the significance of the "aromatic sextet" concept of Eric Clar, whose immense contribution to the chemistry of polycyclic aromatic hydrocarbons is referred to in the introduction to Chapter 30 of the 2nd Edition.

Guide to the Index

This index is constructed in a similar manner to the volume indexes of the first edition of the Chemistry of Carbon Compounds. However, to make the index easier to use, more descriptive entries have been made for the commonly occurring individual, and groups of chemicals.

The indexes cover primarily the chemical compounds mentioned in the text, and also include reactions and techniques, where named, and some sources of chemical compounds such as plant and animal species, oils, etc.

Chemical compounds have been indexed alphabetically under the names used by authors, editing being restricted to ensuring uniformity of entries under the same heading. In view of the alternative nomenclature that can often be used, a limited amount of cross-referencing has been done where it is considered to be helpful, but attention is particularly drawn to Convention 2 below.

For this and the succeeding volumes, the indexing conventions listed below have been adopted.

1. *Alphabetisation*

(a) The following prefixes have not been counted for alphabetising:

n-	*o-*	*as-*	*meso-*	D	*C*
sec-	*m-*	*sym-*	*cis-*	DL	*O-*
tert-	*p-*	*gem-*	*trans-*	L	*N-*
	vic-				*S-*
		lin-			*Bz-*
					Py-

Some prefixes and numbering have been omitted in the index, where they do not usefully contribute to the reference.

(b) The following prefixes have been alphabetised:

Allo	Epi	Neo
Anti	Hetero	Nor
Cyclo	Homo	Pseudo
	Iso	

(c) A letter by letter alphabetical sequence is followed for entries, firstly for the main entry, followed by the descriptive entry. The only exception to this sequence is the placing of plural entries in front of the corresponding individual entries to prevent these being overlooked by a strict alphabetical sequence which could lead to a considerable separation of plural from individual entries. Thus "butanes" will come before *n*-butane, "butenes" before 1-butene, and 2-butene, etc.

2. *Cross references*

In view of the many alternative trivial and systematic names for chemical compounds, the indexes should be searched under any alternative names which may be indicated in the main body of the text. Only a limited amount of cross-referencing has been carried out, where it is considered that it would be helpful to the user.

3. *Esters*

In the case of lower alcohols esters are indexed only under the acid, e.g. propionic methyl ester, not methyl propionate. Ethyl is normally omitted e.g. acetic ester.

4. *Derivatives*

Simple derivatives are not normally indexed if they follow in the same short section of the text.

5. *Collective and plural entries*

In place of "– derivatives" or "– compounds" the plural entry has normally been used. Plural entries have occasionally been used where compiunds of the same name but differing numbering appear in the same section of the text.

6. *Main entries*

The main entry of the more common individual compounds is indicated by heavy type. Multiple entries, such as headings and sub-headings over several pages are shown by "–", e.g., 67–74, 137–139, etc.

INDEX